A PERFECT PLANET

BBC

A PERFECT
PLANET

OUR ONE IN A BILLION
WORLD REVEALED

HUW CORDEY

FOREWORD BY ALASTAIR FOTHERGILL

FOREWORD

I feel very fortunate to have crept within just a few metres of an active volcano. The volcano I visited was not on land but deep beneath the ocean waves. Inside a tiny but extremely strong submersible, we journeyed two miles down to the seafloor to reach the Mid-Atlantic Ridge. It is here, in the midst of the Atlantic Ocean, that two massive continental plates are slowly pulling apart. With every ten metres you descend into the depths, the pressure on the submarine doubles. It had three tiny portholes with extremely thick glass to withstand these enormous pressures.

When, after a two-hour descent, we finally reached the bottom, the pilot turned on the powerful headlights he had been saving to preserve battery life. We started to search the seafloor and, for a good while, we saw nothing but a barren, almost lunar landscape. Then, out of the gloom, emerged tall chimneys, over 20 metres tall, belching out dense clouds of black smoke. This gas was created by volcanic activity just below the surface, and it was extremely hot. We had to be careful not to manoeuvre the submersible too close as the water was heated to up to 463°C. The industrial landscape of smoking chimneys was impressive enough, but the real surprise was the extraordinary variety of life that gathered on the steep walls of the chimneys.

Opposite **The sea arch Gatklettur, on the Snæfellsnes Peninsula of Iceland's west coast, sits on the North American tectonic plate, heading in a westerly direction.**

Below left **In the South Pacific, black smokers spew out dissolved sulphides in super-heated water. The sulphides turn black on contact with cold seawater, forming the 'smoke'.**

Below right **A pink vent fish, a species of deep-sea eelpout, swims amongst giant tubeworms and crabs at a hydrothermal vent field on the East Pacific Rise.**

The black gas that emerges from deep beneath the Earth's crust is rich in hydrogen sulphide. A complete ecosystem has evolved around bacteria that can fix the energy in that hydrogen sulphide. All these animals are found nowhere else on Earth, and they exist without any energy from the sun. Some scientists speculate that life may have first evolved at these hydrothermal vents in the depths of the ocean. What we know for certain is that, without volcanoes, the Earth would be a very different place today.

The forces of nature that dominate our planet have long suffered from a pretty bad reputation. Ever since the eruption of Mount Vesuvius buried the town of Pompeii, we have lived in fear of the destructive force of volcanoes.

Above At dawn, flocks of lesser flamingos feed and preen at Kenya's Lake Bogoria in the East African Rift Valley.

Numerous Hollywood movies have dramatised the tornadoes that rip across the Midwest of the United States. Sailors have long learned to respect the power of the oceans. These fears, though, do not reflect the true story. The forces of nature, when working in balance, have in fact created a perfect planet for life. This book and the television series it accompanies aim to put the record straight. For the first time, here is the full story of how the forces of nature have shaped our home.

Our planet is often called the Goldilocks planet. In the astronomical game of chance we have been extremely lucky. Earth is the perfect distance from the sun – not too cold and not too warm. But the really lucky break was

when a massive asteroid hit our young planet and ripped off material to form our moon. This collision also knocked our planet so that it now lies at an angle of 23.5° with respect to the sun. As Earth makes its annual journey around the sun, it is this angle that creates the seasons, which have been critical to the variety of life we enjoy today. Without it, our Earth would just be two massive ice caps at each pole, two thin stripes of green and a giant desert all around the equator. We are also lucky that our moon is close enough that its gravitational force has locked our planet at the fixed 23.5° angle for millennia.

The moon's gravitational force is also responsible for the daily cycle of tides. Near the coast, the ebb and flow of the tide plays a critical role in refreshing the ecosystem. Combined with the mixing power of waves, the tides deliver nutrients to plants and animals that live in our shallow seas. Further out to sea, in deeper water, currents play a similar role in distributing nutrients. They have also been critical in maintaining a stable climate by distributing hot and cold waters around the planet. Meanwhile, the winds, formed far out to sea, distribute freshwater all around the planet.

Even volcanoes, traditionally seen as the most destructive natural force, have been critical to life on Earth. All the planet's water vapour was originally released by millions of years of volcanic eruptions. Volcanoes were also the original source of all the carbon on our planet and without them there would be no CO_2, no plants, no life. Volcanic activity has also been the architect that

Above left An array of reflective mirrors focuses sunlight onto a solar tower at the Sanlúcar la Mayor thermal energy plant, near Seville in Andalusia, Spain.

Above right Solar cookers are constructed at the Barefoot College in Tilonia, Rajasthan, India.

has shaped our planet. The summits of volcanoes emerging out of the ocean, like the Galápagos Islands, are critical oases for life. East Africa's Rift Valley, the cradle for humanity, was created by two continental shelves that are still pulling slowly apart. Far from being a hostile planet, Earth is a unique and perfect home for the animals and plants that have evolved to live here. For the last 10,000 years, the forces of nature have delivered a stable environment and predictable weather systems. It is this predictability that allowed us to develop agriculture and, ultimately, civilisation.

But, in just the last 50 years, humanity has become the most powerful force of nature. Our ever-increasing influence is threatening the perfect planet, and the evidence is everywhere. In 2019, Europe experienced its hottest June on record. Extreme temperatures in Australia caused the worst ever fires, ravaging 15 million acres, while in the Arctic the extent of the summer ice cap continued to reduce. If we are to find solutions to all these threats, we will need to look to the forces of nature. Every second the sun produces a million times the annual power needs of the United States. We have the technology we need already. With the right political will, we can still preserve a perfect planet for future generations.

Alastair Fothergill

THE SUN

LET THERE BE LIGHT

Just an average-sized star. That's perhaps a harsh description of our glorious sun – the force that, today, nurtures an estimated 8.7 million species on Earth (8.7 million more kinds of life than found on any other planet – at least, so far). Nevertheless, the statement is true. Amongst the two hundred or so billion stars in our galaxy, there are plenty of smaller ones but just as many that are bigger – some hundreds of times bigger.

But the statistics of our sun are still impressive: it has a diameter of nearly 2.3 million kilometres (making it over a hundred times wider than Earth); it's 5,500°C at the surface and 15,000,000°C in the core; and every second the sun produces 400 million, million, million, million watts of power, or a million times the annual power consumption of the USA.

Our sun is composed of two of the simplest elements in the universe: hydrogen and helium. The key is the hydrogen – and the power of gravity. The sun contains over 98 per cent of all the matter in our solar system, and the huge gravitational force acting on this star makes the hydrogen atoms collide with so much force that they literally meld into a new element – helium. This process is known as nuclear fusion. It creates a chain reaction and the result

Below **In a forest clearing in Europe, shafts of sunlight filter down from the crowns of tall beech trees.**

Overleaf **In the setting sun, a humpback whale breaches. It's thought to be one way in which whales communicate.**

is energy, or heat, which, travelling at 300,000 kilometres a second, reaches the Earth in about eight minutes. If that's not impressive enough, the sun has been performing this recurring chemical reaction every second for five billion years and is likely to do so for another five billion.

However, because the sun is very slowly and steadily getting a little denser – a result of converting hydrogen to helium – it's also getting hotter. Over the next billion years, it's thought that the energy Earth gets from the sun will increase by about 10 per cent and, given the current state of our warming planet, we all know what the consequences of that might be. Nevertheless, this is one kind of warming we shouldn't fret about too much. Since the average species on planet Earth lasts between one and ten million years we, as *Homo sapiens*, will be long gone by then.

But of all the facts and figures surrounding our sun, the most important to us by far is its distance from Earth – 150 million kilometres. As it turns out, that is the perfect distance. Any closer, like Venus, and it would be too hot; any further away, like Mars, and it would be too cold. It's why Earth is known as the Goldilocks planet – in other words, just right! The significance of this can be summed up in one word: water. This is the only planet we know of where water can exist in all three forms: vapour, liquid and solid. Life as we know it depends on this.`

THE TILTED PLANET

Another piece of pure cosmic good fortune is the Earth's tilt of 23.5°, the result of a collision with something large over four billion years ago, which scientists sometimes refer to as the Theia Impact. The debris from this collision might then have formed the moon (another lucky break for Earth). Being tilted is not unique in itself as most of the planets in our solar system are tilted to varying degrees – Uranus, for example, has a 97° tilt and Venus one of 177° – but 23.5° turns out to be pretty much the perfect tilt to support life from pole to pole. It's particularly good when combined with Earth's perfect rotational speed of 24 hours. (That's similar to Mars, but Mercury and Venus, by comparison, rotate on their axis every 59 days and 243 days respectively. In other words, a long time in the dark.) Certainly, without any kind of tilt the Earth would be a very different place – most importantly, we wouldn't have the seasons.

Without seasons, the length of day and night would be equal across the planet but, in Earth's northern and southern extremities, the sun would never rise more than a short distance above the horizon, making these zones almost impossible to live in. With no tilt, scientists believe that humans would never have advanced beyond a few scattered communities living around the Earth's mid zones. For a start, we would have struggled to grow one of our most important crops – wheat, which needs cool or cold winters.

Seasons occur because as our tilted Earth orbits the sun every 365 days – or, as we call it, a year – different parts of the planet are pointed towards the sun at different times of the year. In June the Northern Hemisphere is facing the sun, giving it summer, while the Southern Hemisphere is tilted away, resulting in winter. In December, the situation is reversed.

The consequence of this tilt is that, for most of the planet, the number of daylight hours varies from one month to the next. So does the intensity of the sun's energy. What may be surprising is that, averaged out over the year, every square metre of the Earth gets almost the same number of daylight hours – 4,380. It's just delivered in different doses.

Right **A full moon reflects sunlight onto the surface of the Earth, as seen from a satellite orbiting in space.**

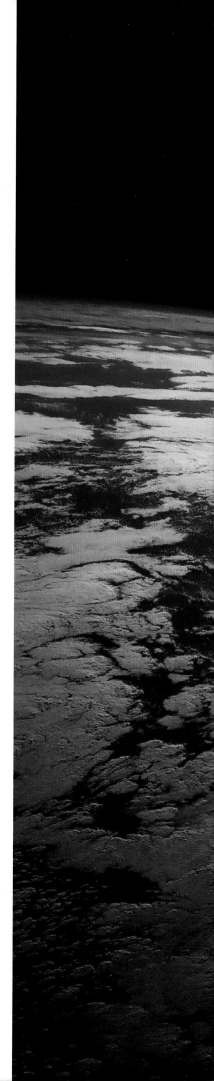

TROPICAL SUN TRAP

The exception to this variation can be found around the equator – home to the planet's tropical forests, jungles or rainforests, where there's a guaranteed 12 hours of daylight every day, all year round. The solar radiation is also greater at the equator because the sun's rays strike the surface more directly (almost at an angle of 90°). This means there is more energy per square metre here than anywhere else. On a bright day, at noon, that's around 1,000 watts per square metre (the equivalent of running a hair dryer for about an hour). It's what makes our jungles the greatest solar power stations on Earth.

It's said that only around 2 per cent of the light hitting an area of tropical forest makes it down to the forest floor, and that's because the structure of this habitat is perfectly designed to maximise the amount of light energy it can absorb. This thirst for light is what drives the trees to heights of 30 metres or more, and why trees don't waste energy producing branches until they reach the canopy and are able to compete with their neighbours. The lack of light on the forest floor is why saplings can remain stunted – as if in suspended animation – for years. But if a light gap opens up through treefall then the sapling turns from a tortoise to a hare. Some jungle trees are born hares. *Cecropia* trees, from Central and South America, are pioneer species quick to exploit light gaps. They grow very fast, reaching heights of 18 metres, but only live for around 30 years – unlike the hardwoods which can go on for hundreds of years.

Every forest leaf is a natural solar panel, each angled to ensure they are working at capacity – though they must also avoid becoming overloaded. At the top of the canopy, the leaves are angled almost vertically so as to catch the early light, while, at the same time, being shaded from the midday sun. Moving down through the canopy, the leaves take on a flatter angle and are more densely packed with light-absorbing cells – again, making the most of the solar radiation. The trees are so efficient at this light-gathering job that they absorb over 90 per cent of the solar energy falling on them – hence the gloomy nature of the forest floor.

The primary reason for all this light capture, of course, is because trees need it to grow (as does every plant and algae on the planet, and many bacteria). They do this through photosynthesis – the process by which millions of chloroplasts inside each leaf take in water and carbon dioxide and turn it into sugars, or food, while, at the same time, releasing that all-important gas, oxygen. What is perhaps more surprising is the fact that only a small fraction of the light hitting a leaf is actually used for photosynthesis – and this is the reason why leaves are green. It's all down to efficiency again. To maximise the amount of energy coming from the different wavelengths of light, each leaf focuses on the red and blue photons (a photon being a light-carrying particle), both of which hold more energy than the green photons. As a result, the green wavelengths are not absorbed by the plant but reflected off the leaf – and into our eyes. So, the colour of a jungle is all down to a leaf's adaptation to light-gathering.

Opposite **The early morning sun breaks through the jungle canopy in Sabah, Borneo.**

Left A brown-throated sloth descends slowly to the lower branches of a *Cecropia* tree in Costa Rica's Manuel Antonio National Park.

The location of our tropical forests is one of the main reasons why they contain as much as 50 per cent of all the species on land, while occupying less than 7 per cent of the world's surface. A uniform climate and reliable quantities of sunlight mean jungles are open for business every day of the year. Here, trees are not forced into a period of dormancy and nothing needs to migrate away, or hibernate. Food is always available somewhere, and one group of trees has taken this to the limit by offering up fruit all year round.

THE GREAT FIG TREE

Fig trees are a keystone species, with more than 600 types across the planet's jungles. They are the only wild trees to fruit all year round and, in the process, they support thousands of different kinds of animals, from gibbons to hornbills, sun bears to bats. They are able to do this because of a partnership with a tiny insect – the one- to two-millimetre-long fig wasp – a relationship that began more than 80 million years ago and, surely, one of the most remarkable pollination stories in the natural world.

Figs don't flower in the conventional way. What we might regard as the fruit is, in fact, an inverted flower. Pollination occurs when a female fig wasp burrows through a tiny opening and into the heart of the fig – something that can only be done on one or two days when the fig is in exactly the right stage of development. This opening is so tight that, as the fig wasp squeezes her way in, her wings and sometimes her antennae are ripped off. But she will never need them again as, when her work is done, she'll die within the fig. Once inside, she lays hundreds of eggs, pollinating the flowers with pollen that she carefully unpacks from pouches in her abdomen. A few weeks later, the closed-off fig nursery comes alive as the eggs start to hatch.

First to emerge are the golden, wingless males, which waste no time mating with their unhatched sisters. To reach them inside the eggs, they use telescopic penises twice their length! When this is complete, they start to dig an escape tunnel to the outside world – not for them but for their now hatching sisters. If this isn't chivalrous enough, these selfless creatures then sacrifice themselves to predatory ants cruising the outside skin of the fig so that their winged sisters can take to the skies.

The female fig wasps now have just a day or two to find another fig in exactly the right stage to burrow into and start the whole process again.

Opposite **A male rhinoceros hornbill feeds on the fruit of a fig tree in Malaysian Borneo.**

Below **A southern yellow-cheeked crested gibbon in Cat Tien National Park, Vietnam. Gibbons spend their lives in the trees and seem to fly through the canopy using a form of locomotion known as brachiating. They can bridge gaps of up to 15 metres in a single leap.**

Opposite **The story of fig pollination.** From left to right: a 2-mm-long female fig wasp pushes her way through a small opening in the fig; a female fig wasp 'swims' through the outer flesh of the fig; laying eggs in the fig's tiny internal flowers; a golden wingless male fig wasp in the act of mating with one of his unhatched sisters; a female fig wasp emerges from her egg case; having dug an escape tunnel for the females, a male emerges on the fig's surface; predatory ants capture the emerging males; a female fig wasp about to fly off. She will live for just 48 hours and in that time must find another fig tree in just the right stage for pollination.

Below **Inside a fig.** Unusually, figs have their flowers on the inside, which is where pollination takes place with the tiny fig wasps.

Despite their tiny size, fig wasps are capable of travelling a staggering 160 miles in 48 hours to find an unripe fig. It's the short life expectancy of these tireless pollinators that is behind the figs' incredible year-round fruiting cycle. If there wasn't another fig in fruit when a new generation of fig wasps emerged, the wasps would die before pollination could take place.

Every single species of fig tree depends on its own kind of fig wasp for pollination and, in case you're wondering, the reason why you don't see a bunch of dead wasps when you bite into a fig is because they are absorbed into the fruit's flesh as it ripens in the sun.

The reliable dose of year-round sun makes a jungle a very stable habitat where there is no need to migrate away, hunker down or hibernate in the face of a sudden and dramatic change in temperature. In these benign conditions, every opportunity is exploited, and every niche catered for. It's why species diversity is so high in these tropical forests and why, for instance, the fig wasp isn't immune from parasites trying to take advantage of what they started.

The fig wasp parasite is another kind of fig wasp and her aim is to lay her eggs in the flowers of those already filled with eggs from ordinary fig wasps. But there's a problem. After a fig wasp enters a fig, the entrance hole is sealed over with sap by the plant so the parasite can't get in the same way. Instead, she must drill through the fig from outside, which she does with a very long ovipositor, or egg-laying tube. It's thinner than a human hair but very tough. An electron micrograph image of the tip has revealed it to be serrated, which is how she is able to pierce the hard, unripe walls of the fig. But, if that's not incredible enough, once she's broken into the nursery with her ovipositor, she has to use it to locate the flowers that are already filled with fig wasp eggs, so that when her larvae hatch they have something to feed on. Tests have shown that the tip of this egg-laying tube has chemical sensors that can detect the flowers with developing eggs, which release carbon dioxide. It's extraordinary and bizarre in equal measure. As one biologist said of the parasite's ovipositor, 'Think of it like a finger with tongues all over it – and now try never thinking of that again.'

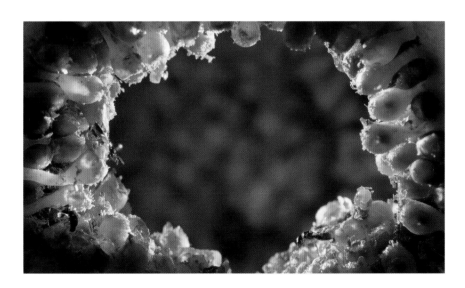

EXTREME DOWNSIZING

Filming any kind of animal is challenging and often frustrating: they might be shy or rare; the story you're after might only take place on a small number of days in difficult-to-reach places; their behaviour might be difficult to follow on a long lens because of the speed of the action; and then there are the vagaries of the weather (and, invariably, the people on location who tell you that you should have been here yesterday / last week / last year because it was *so* much better). But filming a macro sequence has its own unique set of challenges – not least because you are working with a very small depth of field, which makes focusing especially tricky. It's probably why there are relatively few specialist natural history macro camera people out there.

Top of the list for many natural history producers and directors, however, is legendary macro expert Alastair MacEwen. If you've ever seen a macro sequence in a wildlife show and gone, 'Wow!' the chances are that it was filmed by Alastair. Now in his seventies, he's been filming nature across the planet for decades, including sequences for both series of *Planet Earth*, *Blue Planet*, *The Hunt* and *Our Planet*, to name but a few, yet he's lost none of his enthusiasm. He's excited by each new challenge, and the fig wasp story was no exception. Well, that was until he saw the size of his subject.

Even with the remarkable camera technology available to wildlife filmmakers today, capturing the pollination and life cycle of an animal that's just one to two millimetres long is a massive challenge. In the past, it would have been almost impossible because of the kind of film lights we had at our disposal. Filming very small animals requires a lot of light – so as to get enough depth of field to focus – but the old-style lights gave off a lot of heat at the same time, which meant you couldn't get them too close to your subject. But if you had the lights further away, they might not be bright enough for the depth of field you needed. It was a regular conundrum. The new LED lights cameramen now use don't give off heat, however, so that's one problem we no longer have to worry about. But there is still the question of magnification.

Alastair and producer Nick Shoolingin-Jordan discussed the shoot in detail before setting off to film the sequence in Thailand. But talking about filming a tiny insect is one thing – actually seeing a fig wasp in the flesh is another. Get a ruler and mark off two millimetres (and that's a big fig wasp) and you'll see what I mean. It's not a lot larger than the mark a biro makes when you press the point onto a piece of paper. The wasp was certainly small enough to make Alastair gulp, and for both him and Nick to wonder whether they'd bitten off more than they could chew. After all, they weren't just after a few shots of the fig wasp but the entire pollination story, including the moment the wingless males mate with their unhatched sisters. Alastair explained to Nick that what they wanted to film might be too small and beyond the visual possibilities of his lenses.

But they had to try, of course. So, once the field studio was set up, Alastair opened up his macro lens case and produced two very long extension

Opposite **Cameraman Alastair MacEwen with the small microscope lens he needed to film the 2 mm fig wasps. It was the first time he'd used the lens in 30 years of carrying it around in his kit.**

tubes, which, when attached to a macro lens, significantly magnify the image. He got the subject in focus and frowned. It was back to the lens case, and another extension tube came out. This in itself was a rarity but, even with a third tube, it was still not producing the magnification necessary. Alastair and Nick exchanged worried glances.

There's a famous scene in the movie *Jaws* when Chief Brodie is chumming dead fish off the back of the boat and the huge head of the legendary shark (or, in this case, the robotic shark) suddenly rears briefly up out of the water right in front of him. It's his first sight of the shark. Stunned by the size of it, Brodie backs away into the cabin, where he sees Quint, the captain, and utters the immortal line, 'You're gonna need a bigger boat.' Standing by his field set, his brow furrowed with concern, Alastair was about to have his Brodie moment. 'We're going to need a bigger lens,' he said solemnly.

Anyone who has worked with Alastair knows that he carries around a lot of kit and it always includes one Pelican case containing bits and pieces, tools of the trade, specialist items – or to put it another way, a box of *stuff*. I've tried to get Alastair to leave this heavy case behind on a number of trips (or to cull most of the items out of it – not least to save money on excess baggage) but I've never succeeded. And, if truth be known, there's always one moment on a shoot where he'll pull something out of this Pelican case that, as it turns out, we really needed. I'm always amazed at the obscure things that come out of it. Strangely, there seems to be far more in it that you'd think from the size of

Above **Producer Nick Shoolingin-Jordan under a fig tree in Chiang Mai, northern Thailand.**

the case. You could say it's a sort of cameraman's equivalent of Doctor Who's TARDIS. When Alastair does use something from it, he's far too polite to say, 'I told you so.' There's perhaps just a slight raise of an eyebrow and a wry smile.

Anyway, with the fig wasps barely visible in front of Alastair, it was time to reach for *the case*. Rummaging around in it, he eventually found two tiny Zeiss microscope lenses. He told Nick that he had been carrying them around for nearly thirty years but had never used or needed them… until today. After dusting them off, he attached them to the camera, looked down the eyepiece and, after a few moments spent framing and focusing, the monitor lit up with the star of the story, a female fig wasp in all her glory. She wasn't quite filling the screen, but every part of her was in perfect clarity. With these lenses the sequence was now filmable. Over several weeks Alastair and Nick ticked off every element of the behaviour: the female swimming through the flesh of the fig to get to the flowers inside; egg-laying in the tiny fig flower heads; the moment when the female unpacks pollen from pockets in her abdomen; the hatching of both wingless males and females; and the bizarre mating behaviour. As it turned out, this was a little easier to film than some of the other behaviour. A male's penis may be only one-tenth of a millimetre thick, but it is four millimetres long – double the length of the wasp and big enough to show up on even a standard MacEwen lens!

SUNLIT SHALLOWS

Opposite **Scroll coral fluoresces during the night at Bonaire in the Caribbean.**

Above **The tentacles of mushroom coral catch tiny sea creatures on the coast of Indonesia's Komodo Island.**

Overleaf **In the Red Sea, Indo-Pacific sergeant major fish swim along the reef wall of a peninsula at the entrance to the Straits of Tiran, Sharm El Sheikh.**

Coral reefs are often called the rainforests of the oceans because of their extraordinary diversity. Even though they cover less than 1 per cent of the Earth's surface, they are estimated to support around a quarter of all ocean species. And tropical coral reefs and rainforests have another similarity – both rely on regular and reliable sunshine. It's why these kinds of coral reefs are found almost exclusively in shallow tropical and subtropical waters.

The reason behind the spectacular diversity found in coral reefs is a cosy relationship between a coral polyp – an animal – and algae, known collectively as zooxanthellae. Tropical waters are generally low in nutrients but the polyps get around this problem by housing algae within their hard calcium carbonate structure (it's also what gives coral reefs their colourful appearance). The algae depend on sunlight for photosynthesis – hence the need for shallow water – and through this process they provide more than 90 per cent of the energy the polyps require to survive. In return, the algae get a secure home and benefit from the polyps' waste in the form of carbon dioxide and nitrogen.

It's recently been discovered that even deep-sea corals, growing 165 metres below the surface, depend on photosynthesising algae. But how do their algae achieve this feat when, at these depths, there is virtually no sunlight to capture? The answer is fluorescing, which is what causes corals to glow orange or red. Some shallow-water corals produce fluorescent proteins to block the harmful effects of too much solar radiation, which may otherwise damage their algae partners, and it seems their deep-water cousins do the same thing – but for the opposite reason. Fluorescing at this depth boosts the low levels of sunlight by converting blue light, which isn't good for photosynthesis, to orange-red light, which is. These deep-sea algae, in effect, bask in the glow of the corals that house them.

THE LONG POLAR NIGHT

As a result of our tilted planet, the further away you move from the equator the more angled and less focused the sun's rays are. It's like the difference between pointing a torch straight down at your feet and a few metres out in front of you. Straight down, the area being illuminated will be more concentrated, and therefore brighter; facing forward, more of the ground is illuminated but it's less intense or bright. This effect is at its most extreme at the poles, which is why, even in midsummer, the solar energy hitting the ground there is much less per square metre than at the equator. And if the ground isn't as hot, then the air above it isn't as warm, and that's another reason it's colder in the polar regions even when the sun's out. But what must it be like when the sun doesn't make an appearance for months at a time?

Ellesmere Island, in Canada's High Arctic, is 196,000 square kilometres, making it the tenth-biggest island in the world. It's just a shade smaller than Great Britain but with a vastly smaller population; in fact, fewer than 200 people live in Ellesmere. The low population density isn't surprising: temperatures in winter regularly drop to minus 50°C and for four months – from the end of October to the end of February, when the Northern Hemisphere is tilted away from the sun on the Earth's annual orbit – the sun never rises above the horizon. It's easy to understand why Ellesmere Island is considered to be one of the coldest inhabited places on Earth.

In terms of extremes, Ellesmere Island is right up there amongst the world's most hostile spots. Indeed, a combination of Ellesmere's harsh environment and alien landscape is why NASA chose it as a location to run Mars simulations – since it's probably the place on Earth that most closely resembles our planetary neighbour. As cameraman Kieran O'Donovan poetically put it during our shoot, 'There is a terrifying purity to the emptiness and silence of this place and everything about it feels a barrier to even just the idea of life. I feel like a terrestrial astronaut exploring Canada's far north – looking every bit the part in the cumbersome winter "spacesuit" I have to wear to shield me from the hostile atmosphere. In these temperatures,

Opposite **The full moon casts an ethereal glow across Ellesmere Island in the Canadian Arctic. For several months during the winter, the sun never rises above the horizon.**

Below **The frozen wilderness of Ellesmere Island in the Canadian territory of Nunavut.**

frostbite can happen in just minutes so nothing can be left exposed.' (Even with these precautions, Kieran got frostbite on his cheek, which meant staying out of the elements for a few days.)

During Ellesmere's winter, the only light comes from the moon – or, to be more precise, sunlight from the other side of the planet bouncing off the moon, which creates an otherworldly beauty. As Kieran said, 'When it's not in complete blackness, the light of this world comes in countless shades of soft alien blues, purples and pinks – smoothly graded from horizon to landscape.'

Despite Ellesmere Island's huge challenges, a few species do live here all year round and are even active during the long months of darkness.

THE WOLF AND THE OXEN

Arctic wolves – smaller relatives of the grey wolf – prowl this moonlit landscape, ghostlike in their white coats. Ellesmere's wolves have had a continuous presence on the island since the last ice age 14,000 years ago. So, not surprisingly, they are well adapted to what we would regard as mind-numbingly low temperatures. A double layer of fur insulates them from the biting wind; smaller, rounder ears and shorter muzzles than grey wolves result in a lower surface area to volume ratio, which prevents heat loss; hair between the pads of their toes helps to keep their feet warm and gives them more grip on icy surfaces; and, to stop cold from the freezing ground permeating upwards and cooling their cores, they are able to regulate the temperature of their foot skin to 0°C – even when the substrates they are standing on are at minus 50°C. Although they may be dressed for the climate, however, life for Ellesmere's white wolves is no picnic.

The wolves hunt throughout the dark winter months – and especially when moonlit periods offer more light to see by. At a time when conserving energy may seem a good strategy, these predators often have to travel large distances – as much as 40 miles a day – in their search for food. In winter, their choices are few but at the top of Ellesmere's very small menu for wolves are muskoxen.

Muskoxen are remnants of the ice age – with the looks to go with it, particularly after a storm when snow sticks to their faces, giving them icy masks. These herd animals seem to fit this hostile landscape perfectly, weighing up to 410 kilograms, with a long shaggy coat and thick bony forehead, which flows down to a set of sharp curly horns (which, observed head on, look a bit like an old-fashioned Dutch bonnet). And like the wolves, they are very well adapted to the cold.

The muskoxen's principal defence against the sub-zero conditions is their unique double-layered coat. The top layer is made up of long, coarse hair, which keeps snow off its body. The layer beneath – known as qiviut – is made of short, dense hair which traps warm air coming from the body of the animal. It's this that insulates it from the extreme cold. This layer is thought to be the warmest animal coat in the world, eight to ten times warmer than that of a sheep. It's so effective that even in freezing, gale-force winds, when other Arctic animals have to hunker down, out of the elements, the cold air never reaches their skin. But if conditions get *really* bad then the muskoxen can always huddle together for added warmth.

Winter offers very meagre rations for these large herbivores. What little there is, such as mosses and lichens, is buried under a blanket of snow – so, to get at this food the snow must be scraped away by the adult's hard, sharp hooves. This winter fuel is just enough to take the edge off the hunger, but at this time of the year the muskoxen must depend on fat laid down during the summer months, when the sun's energy powered plant growth. To make their reserves go further in winter, they reduce their energy consumption by being less active and slowing their metabolic rate. Muskoxen can, in fact, reduce

Opposite **A white or Arctic wolf – its face dusted with ice and stained with the blood from a recent kill, Ellesmere Island.**

their metabolic rate by 30 per cent during winter: a very useful adaptation when there's so little food around. In these hardened times of the year, the last thing the muskoxen want is to be hassled and chased by wolves.

Muskoxen may be the number one food choice for Ellesmere's wolves but, nonetheless, they are formidable quarry. The only way the wolves can hope to get the better of one is to work together in a pack. The key is to separate one muskox from the herd, and they do this by running at the group. It's not an easy win as the muskoxen's natural instinct is to stick together and form a tight defensive huddle – heads and horns facing outwards, so protecting each other and the more vulnerable youngsters in the middle. If the wolves are lucky, the whole herd might spook and run. For Kieran, one such occasion resulted in a powerful and memorable sight: 'On a day when the ambient temperature was minus 46°C, I watched the wolves chase a herd of muskoxen through air so heavy that the contrail of expired respiration and sweat coming off the muskoxen's panicked bodies hovered in their wake, leaving a sinuous line of fog gently drifting over the snow, long after the chase ended.'

When a herd is circled by hungry wolves, there's clearly a strong urge for some muskoxen to charge – after all, one on one, a wolf must seem an insignificant threat to a muskox. But that is precisely what the wolves are hoping for. If a muskox becomes separated from the crowd, the wolves immediately surround it, taking it in turns to attack its rear, while keeping out of the way of the muskox's lethal horns. It's a battle of attrition for both parties – with both the muskox and wolves spinning around each other in an effort to get the upper hand. If the wolves come out on top, the carcass may last the pack a week or more. But even when the wolves appear to have the advantage, the tables can sometimes be turned by the rest of the herd returning en masse to defend their isolated member and bring it back into the fold. If the muskoxen succeed in closing their ranks again, it's highly unlikely the wolves will get a second chance and any effort they've invested will have been wasted.

Above **Cameraman Rolf Steinmann films an inquisitive white wolf. These wolves see humans so rarely that the crew were a curiosity.**

Opposite **A pack of white wolves circle a herd of muskoxen, who form a defensive ring. The wolves' strategy is to separate one from the herd.**

HERD ON THE HOP

During the winter, the wolves' only alternative food source is Arctic hares. Like the wolves, Ellesmere's hares stay white all year round – the short summers of the High Arctic make it an inefficient use of energy to flip the colour of their fur between white and brown, as some other Arctic animals do. The hares may stand out like sore thumbs in summer, but their snow-coloured coats are great camouflage during the winter.

In Ellesmere's moonlit months, the hares can sometimes be seen in groups of hundreds. Nobody is quite sure why they do this, but it could be for safety – while some are digging for mosses and lichens, others are keeping an eye out for danger. Or it could be that by huddling together in numbers they may also be getting protection from the elements. Either way, if a pack of wolves comes across a herd of hares it's an irresistible attraction – a buffet on the hop. But while the wolves employ a strategy of cooperation for a

Opposite **During the winter, Arctic hares gather in groups of hundreds – behaviour that's never been filmed before.**

Below **A white wolf attempts, unsuccessfully, to catch an Arctic hare.**

muskoxen hunt, when it comes to hares it seems to be every wolf for itself – though again there is no guarantee of success. Trying to pick one white hare out of a herd of hundreds – all moving and jinking fast across the snow – is clearly very challenging. Indeed, the hares appear to have the upper hand in these winter contests.

When the sun finally appears above the horizon at the end of February, it's a relief for all of Ellesmere's residents. On the first day, the sun rises for a little less than 25 minutes but in under two months the number of daylight hours will hit 24. By the time that happens, the wolves will be able to focus their attention on new-born muskoxen, as well as a crop of less savvy hare babies, or leverets. The muskoxen and hares, on the other hand, will be able to fatten up on summer's abundant vegetation.

CRYOGENIC FROG

For many animals, the answer to surviving a sudden drop in temperature, brought on by the effects of our planet's tilt, is to hibernate. This is defined as a state of suspended animation where an animal's body function slows down in order to conserve energy and cope with seasonal challenges. Typically, this will be characterised by a decrease in body temperature, heart rate (down to as little as 2.5 per cent of the usual levels) and breathing. Some animals actually stop breathing completely! There isn't just one kind of hibernation but a sliding scale from 'true hibernators', like marmots, to a 'deep sleep', such as bears, and finally an 'occasional sleep', which raccoons do. But of all those going into a form of suspended animation, surely the most extraordinary is the strategy one species of frog uses to get through the tough months of winter in the Northern Hemisphere.

Hibernation is regulated by temperature – not just at the start of the process but also at the end – so when the air warms up the animal wakes up. In the case of the humble wood frog, waking up can't happen until it first defrosts. While in its hibernaculum – a shallow burrow on the forest floor – it literally freezes solid like a block of ice, and it remains in this cryogenic state all winter. The frog becomes hard and crunchy and, according to one biologist, 'it goes "clink" when you drop it'. Freezing can cause irreparable damage to healthy

Opposite **A wood frog, none the worse for wear after several months frozen solid, Ohio, USA.**

Below **As spring takes hold in North America, a wood frog slowly returns to life. Since the start of winter, it has been frozen solid like a block of ice. Its heart stopped beating and its blood was frozen.**

cells (resulting, for instance, in frostbite) but the frog avoids this through a process that first sucks most of the water out of its cells and then fills each with a sugary syrup. This sugar solution acts like a natural antifreeze.

But, when the sun's power is turned up in spring, something remarkable happens: the frog's blood melts and begins to flow, and its heart starts to beat again. In just seven hours, it, almost magically, thaws back to life: a defrosting trick that means it's primed and ready to go the moment spring arrives.

A DEN OF SNAKES

The spring emergence of one reptile, in Canada's Manitoba province, is equally as dramatic as the wood frog – though in a very different way. For several days in May, when the sun's rays permeate down into the bedrock, red-sided garter snakes emerge from their winter dens – natural caverns formed in the limestone rock. It's not just a handful of snakes, or even a few hundred, but tens of thousands, making this event the greatest gathering of snakes on the planet.

As with all hibernating animals, it's the air temperature that rouses the garter snakes from their sleep but when they emerge into the open there could still be snow on the ground, which makes for an unusual juxtaposition – the sight of a cold-blooded reptile sunbathing on snow. The males are the first to appear. After months of hibernation, they are slow, weak, cold to the bone and in desperate need of the sun's energy. If the temperature continues to rise, they will remain above ground, warming up in the sun. If there's a sudden cold snap, they'll be forced back down to the dens again – though rarely for long. With the seasonal clock ticking, the snakes know they must take every opportunity to get their lives back up and running. There's a lot to get done in a short space of time and, though they won't have eaten since October, right now, the males have just one thing on their minds – females.

Shots of thousands of writhing snakes are striking enough but seeing them in the flesh here is something else. It's like being on an Indiana Jones movie set. While filming the emergence, you can literally become surrounded by snakes as they crawl around and even over you. Garter snakes are non-venomous, and fortunately they're also entirely unaggressive so, despite their numbers, there's really nothing to worry about – at least at this stage of the year when the snakes are totally absorbed with the idea of mating. (I imagine if you tried to pick one up in July it would attempt to bite you.)

Opposite **In one of the greatest gatherings of reptiles on the planet, thousands of garter snakes emerge from hibernation in Manitoba, Canada.**

Below **Male garter snakes emerge first and once warmed by the sun are ready to mate with the much larger females. Males outnumber females by a hundred to one, so competition for each female is intense.**

Overleaf **Garter snake on ice, Manitoba. With the seasonal clock ticking, garter snakes emerge from their underground dens at the first sign of spring – even if there is still snow on the ground.**

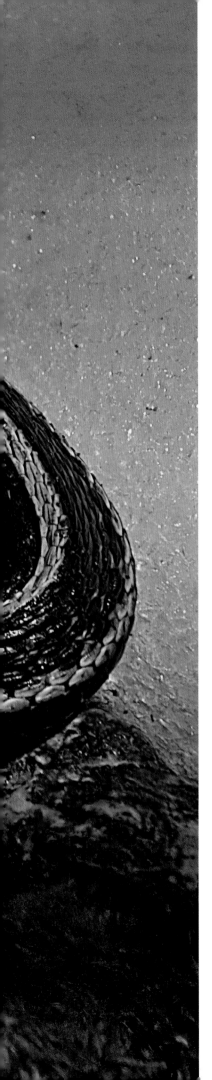

Nevertheless, if you have a phobia about snakes the Narcisse dens in Manitoba are quite definitely the one place you should never go to in May. So it was a big surprise, during a previous visit to the dens on another project, when the natural sounds of spring in northern Canada were broken by what can only be described as a bloodcurdling human scream. We nearly jumped out of our skins, which seems like an odd thing to say when you're completely surrounded by snakes and you're not referring to the thousands of gyrating reptiles in front of you. Looking up to the edge of the snake pit, we saw a Japanese woman in a state of terror, with a couple of people at her side. Close by, a cameraman was capturing the whole drama of her reaction. After a few minutes the shaking, and still distraught, woman was led away. It turned out she had a massive fear of snakes and was being filmed for a Japanese TV show on phobias. Apparently, the producers hadn't told her exactly what she was going to see and had even blindfolded her on the walk to the dens so they could capture the exact moment of her worst nightmare.

Fortunately, male garter snakes are so focused on the first job of the year – mating – that not even a screaming human will put them off their stride. Once they have warmed up, they're ready for the much larger females, which now begin to emerge, fashionably late to the party. The males outnumber the females by a hundred to one, so competition for each female is intense. And, if the males aren't hyped up enough, they're stirred into an even greater frenzy of excitement by a female's first act after emergence – the release of an alluring pheromone. She must warm up too, but her strategy is to use the males' body heat to speed up the process. It doesn't take long. Wrapping themselves around her and vying for attention, there might literally be dozens of males trying to seduce one female – numbers that make mating near impossible. So the female has a test to separate the men from the boys. She makes an ascent of the snake pit walls and whoever is still clinging to her by the top wins her hand – or the snake equivalent. There might be several suitors that manage to get across the finishing line, in which case she will mate with all of them. Once all the females have emerged and been mated, each snake heads off across the grass. It will be a solitary life until falling temperatures drive them back to the dens in October.

There's one interesting footnote to the story of the mating garter snakes. It seems there are some males that try and cheat the system and, rather than rely on the power of the sun to slowly warm them up, get other males to do the job for them. These 'she-males', as scientists call them, produce pheromones that make them smell like females. It's too strong to ignore, and the waiting, sex-crazed males move straight to stage one of courtship. They wrap themselves around the sneaky she-males – transferring their hard-won body heat to the cheat in the process. She-males only stay in this state for a day or two but that's all they need to get what they want. And why do they want it? Well, researchers believe that warming up quickly makes them less vulnerable to predators like crows.

THE MIDNIGHT SUN

Below A feisty mother snow goose mobs an Arctic fox intent on stealing her eggs on Wrangel Island in Arctic Russia.

In the high latitudes of the Northern Hemisphere, summer doesn't last long, but what it lacks in quantity of weeks it makes up for in quantity of hours. In places like Karrak Lake in Nunavut, Canada, summer is a round-the-clock affair, with non-stop daylight. This short but intense growing season provides such a bounty of fresh vegetation that it attracts visitors from more than 1,500 kilometres away to the south.

In just a week or two, Karrak Lake can go from a dead, featureless expanse of snow and ice – where the only sound comes from the howling wind – to a bustling metropolis of nesting birds. Even for experienced field biologists, this transformation never ceases to surprise. As Alain Lusignan, our field and camera assistant, who's spent six summers in Karrak, said, 'The sudden change in the soundscape is unmistakable. Stepping out of the bunkhouse in the morning, instead of silence, you are met with the sound of squawks and honks filling the air from horizon to horizon.' The snow geese have touched down. Over a couple of weeks, goose numbers will rise to almost a million, making this one of the greatest gathering of birds on Earth.

Snow geese mate for life and each pair travels together on their migration north. If they've timed their arrival right, they can get straight on with the business of building a nest, but Arctic weather can be unpredictable. Arrive too early, before the snow has melted, and they won't be able to start nesting; too late and they may lose out on the choicest nesting spots. The best sites are in the middle of the colony and there's fierce competition between couples for this high-end real estate. Losers will be forced to nest on the outskirts, where they will be more vulnerable to predators like Arctic foxes. Not surprisingly, the densest part of the colony is the most impressive, and here white geese dot the landscape as far as the eye can see. According to Alain, a person could hike for hours and still not exit the colony.

Once the geese have nested, the Arctic foxes go into action. They've spent the year in Karrak, and the arrival of the geese is a lifeline after more than six months eking out an existence on the frozen tundra. For breeding foxes, it's even more vital as they time the birth of their cubs to this seasonal influx. By the time the pups are strong enough to emerge from the den, they will each need 300 calories a day. Multiply that by, say, ten (and Arctic foxes can have up to a staggering 26 pups) and that's a lot of food to find. The parents will have to work around the clock to provide, and not just for their young. This is also their big opportunity to cache food for the long months of winter.

As soon as the geese start incubating their clutch, their eggs become top of the fox menu. To get at them, a fox must drive off both the brooding parent

Above **Arctic foxes are a constant threat to nesting snow geese. Here, one is successful in chasing the goose from her nest on Wrangel Island.**

Above **With the mother snow goose gone, the fox steals one of her eggs. It might collect up to 40 eggs in a day and put them in cold storage, to be eaten later, when other foods, such as lemmings, are unavailable.**

Overleaf **Arctic fox mother with pups. Both parents share the duty of raising their offspring – a tall order when litters can exceed 20.**

and its partner standing guard. The birds are no pushover, however, and few pairs give way without a fight. The would-be egg thief is met with jabbing beaks, flapping wings and a full-on charge. A vigorous and sustained defence may win the day for the geese – though not often. The foxes of Karrak Lake have a knack for thievery and are very opportunistic. Occasionally, a fox can pilfer a nest unchallenged, if it's spotted the moment when a pair of geese leave the nest briefly to get food and water.

In just three weeks, one fox may steal over 800 eggs. Some are food for the hungry pups, interspersed with good helpings of lemming, but many of the eggs are buried in the ground – vital food to survive the winter. Some foxes clearly take this more seriously than others. In one study, an Arctic fox was found to have cached 136 sea birds in a single larder.

When the eggs have hatched, the foxes move on to goslings (caching these too), even snatching the occasional adult goose. For the foxes, it's all about making hay while the sun shines – though smash and grab may be a more accurate portrayal. From previous years, they know the colony will soon be gone. Indeed, almost as soon as the goslings hatch, the parents begin to march them northwards to the coast, where the grass is more nutritious and abundant. By the middle of July, the colony is empty. All that's left are abandoned nest bowls and downy feathers scattering in the wind. The once bustling colony now has the atmosphere of the morning after the end of a music festival.

GOOSE LULLABY

Filming at Karrak Lake has many challenges and it's not just the sheer scale of the place, which is part of the Queen Maud sanctuary – covering 62,000 square kilometres, it's the largest protected area in Canada. Once the snow has fully melted, in the early weeks of June, getting around becomes much more difficult. Skidoos must be swapped for boats as Karrak Lake now floods and is transformed into a maze of shifting ice pans. What might have been a clear passage in the morning can close up a few hours later with a shift in wind direction. 'Often, there's no choice but to drag one's boat across the ice, like some early Arctic explorer,' said Alain.

Then, there's the issue of deciding which part of the 24-hour day one films in and adjusting body clocks accordingly. The activity of animals living in non-stop daylight varies from species to species: some rest during the night (or what would be the night), others during the day, and some are active all the time, though there may still be peaks and troughs in this activity.

Arctic foxes are active most of the time, but they're *more* active in the night window, and since the light for filming is better then (a result of the sun being lower in the sky), the crew settled into the same nocturnal rhythm as the foxes. Karrak Lake's researchers, on the other hand, worked the daytime shift so it was only at dinner that crew and scientists met.

Below Cameraman Ivo Nörenberg filming snow geese and Arctic foxes in Karrak Lake, Canada. Even in the middle of summer, when there are 24 hours of daylight, it's hat and coat weather on the tundra.

The main aim of the shoot was to film Karrak's foxes stealing and caching eggs. The challenge, as with many wildlife film shoots, was being in the right place at the right time – and predicting that wasn't easy given that each fox had literally thousands of nest options to choose from. So, as is often the case, the best strategy was to pick the most likely spot, based on field observations, and hope for the best. Fortunately, perseverance paid off – with help from the chilly temperatures, when the sun was at its lowest, which kept cameraman Ivo Nörenberg alert. During the long hours spent staking out the colony, Ivo said, 'If it wasn't for the cold, the constant chatter of geese was almost enough to sing you to sleep.' But while Ivo might have struggled to stay awake at times while listening to the goose-call lullaby, he was certainly never bored. For him, as for most wildlife camera people, true wildernesses like Karrak are places to treasure. Today, there are fewer and fewer locations where you can scan a full 360° and never see a single human structure. 'I just can't get enough of that,' said Ivo.

NATURE'S RHYTHMS

All plants and animals have a built-in biological clock – known scientifically as circadian rhythms – which is approximately synchronised to the 24-hour rotational cycle of the Earth. Put simply, it's the body's wake / sleep cycle. The sleep part is driven by a hormone called melatonin, which is produced in the pineal gland in the brain. For diurnal animals, like humans, melatonin production starts in the evening, reaching a peak in the middle of the night, and stops at around 5 or 6 a.m. – it's what makes us tired at night and alert in the morning (or most of us anyway). It's also the reason for jet lag when crossing time zones too quickly. This body clock doesn't completely depend on light as other things influence it too, such as mealtimes and temperature, but without light an individual's circadian rhythm can become seriously unbalanced, affecting mental and physical health. Daylight cycles are the most effective way of resynchronising the biological clock.

As well as a 24-hour clock for feeling alert and tired at the right times, animals living in more temperate places also need a longer-term internal calendar so they can effectively 'predict' the future and anticipate seasonal conditions. Known as photoperiodism, it's the ability of plants and animals to measure day length to determine the time of year. Without this knowledge, animals wouldn't know the best times to carve out territories, mate, have their young, or even migrate. Having a baby, for instance, in midwinter when there's little food around would, in most cases, be pretty disastrous. Likewise, male songbirds would be wasting a lot of energy if they put all their effort into singing when females weren't interested in breeding.

The impact this continuous body calendar has on some species' physiology and behaviour is more surprising than you might think. The annual shortening and lengthening of daylight even causes a change in the eye colour of reindeer, from gold to blue. Gold is better during the endless daylight of summer, while blue is more light-sensitive, which helps the reindeer see more in the gloom of winter. The reindeer's internal calendar also leads to a seasonal shift in melatonin. In summer, when there's much more food around, melatonin production is suppressed so the reindeer doesn't feel tired, meaning it can graze almost continuously.

Photoperiodism also makes the brains and sexual organs of some animals shrink in the winter so that more of their resources can be channelled into functions like thermoregulation. It can lead to an increase or decrease in aggression – increasing when it's time to carve out territories and mate (usually at a time when there's more food around), but decreasing in winter when it's more important to conserve energy. In animals that have their young in early spring but have a relatively long gestation period, like bighorn sheep, their internal calendar will tell them to mate as the days get shorter (so they're known scientifically as 'short-day breeders'). For long-day breeders, like Siberian hamsters, which have a very short gestation, photoperiodism generally dictates that mating occurs during late spring and early summer.

Opposite **A mother bighorn sheep and her lamb on a hillside in western Alberta, Canada.**

Overleaf **A male red-winged blackbird sings at sunrise on a cold morning in Davis, California.**

Right A female reindeer and her twin calves at Svalbard, Norway. The Svalbard reindeer is a small subspecies that has lived in the Svalbard Archipelago for over 5,000 years and so is well adapted to the icy conditions.

GREEN TO GOLD

Photoperiodism is also the reason why leaves change colour at the end of summer's growing season. As the days shorten, the reduction in sunlight makes photosynthesis no longer worth the effort. As a result, deciduous trees start to shut down their leafy solar panels, in preparation for winter, and they

Below **Autumn (fall)**
colours dominate the
landscape at Crested
Butte, Colorado, USA.

know when to do this because they, too, are measuring day length. If they didn't take this action, water within the leaves' cells would freeze and kill the leaves anyway, so taking control of the process saves the trees vital energy.

During this period, chlorophyll, vital for capturing sunlight during photosynthesis, starts to break down, though some of it is absorbed back into the tree. Without the green chlorophyll, other natural pigments, like carotenoids and xanthophylls, which are always present in the leaves, begin

to show through. It's this that makes the leaves flush with red, gold, yellow and brown, and marks the arrival of autumn or fall. It's one of the greatest transformations in the natural world – so big, in fact, that in some regions, like New England in the USA, it can be seen from space.

At the same time, the tree starts to cut its connections with each leaf by forming waxy and impermeable cells at the join. This acts as a protective seal for when the leaf falls. When the process is complete, the leafless tree enters a dormant state – effectively hibernating for the winter.

In the temperate regions of our planet, the seasons are like a natural traffic light for wildlife. The green of summer means *go* – make the most of it. The orange and red of autumn give a visual cue that time is running out and, for animals like snub-nosed monkeys in China's Sichuan forest, that means feasting on as many pinecones as they can get before winter comes roaring in. Once the cones have gone, there will be very slim pickings for these hardy monkeys until the green returns.

Above **With winter approaching, two male snub-nosed monkeys fight over dwindling resources, Qinling Mountains, Shaanxi Province, China.**

Opposite **A troop of snub-nosed monkeys can contain several hundred individuals but group size fluctuates throughout the year. The monkeys are, perhaps, more commonly seen in smaller social units of between 9 and 18 individuals, made up of a single male with multiple females and their young.**

SOLAR TECH

The hottest places on the planet are not on the equator, even though this strip gets the highest concentration of sunlight every year, but in the tropical zones north and south of the equator, where rainfall is less than 25 centimetres a year – the accepted definition of a desert. The largest of these is the Sahara, which is scoured by warm, dry air that originally rose over the equator – though on its journey from the jungle the air has become as much as 10°C hotter.

With very low rainfall, vegetation is naturally sparse and, in the absence of clouds and tree cover, the ground is subject to intense solar radiation. Animals living here must, of course, be adapted to life with little to no water but they also need strategies for keeping *out* of the heat – particularly during the middle of the day, when air temperatures can rise to over 50°C and ground temperatures push 65°C. Get caught out at this time of the day and it's likely to be game over. Unless you're a silver ant.

It's not just mad dogs and Englishmen that go out in the midday sun. Noon is precisely when this record-breaking ant emerges above ground to scavenge the sand for sun-scorched victims. It might seem reckless in the extreme, but being active now means they have nothing to fear from predators, who are all sensibly hiding from the sun. And because literally nothing else is crazy enough to be out and about on the sand at this time, they have no competition for any insect carcasses that they find.

To survive the intense heat, the ants are equipped with solar tech to help stop their bodies overheating. Each is covered in special glass-like hairs that reflect the sun's rays – and are the reason for their silvery colour. Silver ants are also the fastest ants in the world – and perhaps the fastest animal on Earth relative to their size – able to move a staggering 85 centimetres a second. Recently, scientists have discovered that, when the silver ants run at this speed, they can almost fly through the air with all six legs off the ground at once: a very useful adaptation when the substrate is hot enough to fry an egg. Nevertheless, despite these special qualities, the silver ant is still only able to spend a few minutes in these temperatures. Any longer and it'll succumb to heat stroke, like the food it's after. So, getting lost out on the dunes would be a disaster. To avoid this, the ant spins around every few seconds to take a bearing from the sun. When its time is up, it can then make a beeline back to the den and safety. If an ant finds a victim, it will drag it back to its underground home. If the food prize is too big, others will need to assist in the transport, or carve it up into smaller pieces. However, butchery takes time, so the ants need to weigh up the risk versus reward of doing this – particularly if they're already some distance from the nest. Even for one of the world's greatest solar specialists, seconds count.

Opposite **Saharan silver ants, southern Morocco.** These ants are equipped with solar tech – reflective hairs and long legs – allowing them to withstand ground temperatures of up to 65°C, at least for a short time. These ants have found a sun-scorched beetle and are in the process of dragging it back to their underground den.

TOO HOT TO HANDLE

Heat – or too much of it – is obviously a major issue for people too. The saying about mad dogs and Englishmen may have some foundation, but the human body is physically incapable of cooling down quickly enough when the mercury moves past 50°C. In fact, before human inventions like air-conditioning, it was almost impossible to live full-time in some places around the world. So, how do you film a sequence which all takes place at that temperature and where air-conditioning isn't an option? The simple answer is: with great difficulty. Indeed, it's a challenge not just for a film crew but for the camera kit too, as Nick Shoolingin-Jordan, the director, found out on day one of the shoot in southern Morocco. Nick said the air temperature was so high that the camera's cooling fan was sucking in air that was actually heating up the camera. And the result of this was that the camera shut down, repeatedly. This was closely followed by Richard Kirby, the cameraman. As Nick said, 'Despite cold drinks, umbrellas and a fierce determination, as Richard lay on the scorching sand he was in serious danger of overheating.' Nick needed a solution for both, and fast.

Lots of 'technical' things were tried without success – like creating a tin-foil heat sink for the camera (which kept blowing away with the gale-force desert wind). The eventual answer came from the team's local fixer and was surprisingly low-tech (though field solutions often are, particularly in remote locations). Two large pieces of snow-white cotton material, normally used for turbans, were soaked in water and then wrapped around both Richard and the camera. In simple terms, the aim was to use the hot desert wind to cool both through evaporation – or, as it's known in physics, latent heat. This technique dropped the temperature by 15°C, just enough to allow cameraman and camera to continue in filming mode – though of course the cotton turbans needed to be regularly re-wetted.

And the heat wasn't the shoot's only issue – just as taxing was trying to film small critters that ran almost as fast as the grains of sand spinning across the dunes on the hot desert wind. In fact, adjusting for scale, Usain Bolt would have to run at 760 kilometres per hour to reach the equivalent speed of a silver ant. Following and focusing on super-fast ants while wrapped in a sheet of wet cotton wasn't easy, to put it mildly, but after several weeks Richard and Nick finally captured the sequence they'd set out to get. According to Nick, part of the success was down to the small bar they found on the edge of the desert, where after a day toiling in the midday sun they slaked their thirst with a beer and regained their sanity.

Opposite **Cameraman Richard Kirby filming Saharan silver ants in southern Morocco. Wrapping his camera, and often himself, in a length of wet, white cotton fabric was the only way to keep both kit and himself going in the blistering daytime temperatures.**

A LIFE IN THE SUN

The idea of endless summer sun would be a dream come true for many people – particularly those living at higher latitudes, where a change of season brings long nights and short days that may be cold, damp and grey. But for some animals that is an achievable goal.

Sooty shearwaters – one of the world's most abundant seabirds – cheat the seasonal changes the Earth's tilt brings by moving north and south. Every year, they clock up a staggering 64,000 kilometres – equivalent to going one-and-a-half times around the world – as they journey from the Southern to the Northern Hemisphere and back. In a lifetime, they might travel over 1.6 million kilometres – that's the distance to the moon and back. It's the longest migration in a single season by any species – a fact confirmed by satellite tagging data. Arctic terns travel further as they migrate between the Arctic and Antarctic, and will often fly thousands of miles out of their way to take advantage of the best weather and food. But, as yet, scientists haven't confirmed whether they do it in just one year.

Below An inveterate migratory species, the sooty shearwater hunts close to the sea's surface off the coast of Victoria, British Columbia, Canada. It breeds in the Southern Hemisphere.

It's hard to say whether sooty shearwaters are Southern Hemisphere birds that go north or Northern Hemisphere birds that go south, but they nest in the deep south, like on Snare's Island, off the bottom of New Zealand. (Breeding isn't necessarily the key to understanding a bird's nationality as Swainson's hawks, for example, migrate between North and South America every year but are considered to be South American birds that go north to breed.)

Shearwaters feed on fish, squid and krill, the numbers of which increase or decrease depending on the time of the year. But all are at their scarcest during the winter – and that is what drives the sooty shearwaters on their annual pilgrimages north and south. There are a number of reasons why animals migrate, but food availability is perhaps the most common.

The seasonal peaks and troughs of food can obviously have a dramatic impact on the local wildlife. Take the blue tit, one of our most common garden birds. They are capable of living to the ripe old age of 21, but in the UK their average life expectancy is under three years (2.7 to be exact). Compare that to a similar-sized bird in the tropics, such as the lance-tailed manakin, whose maximum lifespan is roughly the same as the blue tit but whose average life

expectancy is over 12 years. The reason for this difference is that the manakin has a regular year-round supply of food and lives in a part of the world with reliably warm temperatures. The blue tit, on the other hand, has to cope with a sudden and drastic shortage of food, while facing freezing conditions at the same time. It's a wonder why any animal hangs around when winter starts to bite. But the ability of shearwaters and Arctic terns to migrate tens of thousands of kilometres has evolved over millions of years, which is why there are comparatively few examples of species performing similar feats of endurance. To slightly misquote the lyrics from a well-known song, 'When the going gets tough, only the toughest can get going.'

So when the amount of food starts to drop off in the Southern Hemisphere, the shearwaters hitch up their well-fed bellies and launch off across the Pacific. Riding wind currents where possible, their epic journey north takes them to one of three places: Japan, Kamchatka or the Aleutians, which is a chain of

Below **Having journeyed for thousands of kilometres, sooty shearwaters and humpback whales make the most of the summer's bounty of krill off the Aleutian Islands, Alaska.**

islands running west from Alaska. Nobody knows what determines the choice of location, but strangely even siblings from the same nest can end up going to different places. But no matter which of the northern summer spots each bird picks, the key is to arrive just as sun-fuelled plankton blooms send the production of their favourite food – krill – into overdrive.

In the Aleutians, the shearwaters are not the only animals to capitalise on the seasonal bounty. Each year, up to 6,000 humpback whales travel north from their breeding grounds in the tropics, where they've barely eaten for six months, to gorge on krill. The humpbacks attack the shoal from below, the shearwaters from above, though these birds can dive to depths of nearly 70 metres. Together, humpbacks and shearwaters create one of the greatest gatherings of life on Earth, and it's only made possible because these animals have found a way to live in what seems like an endless summer.

WEATHER

RAINING AGAIN?

We British are well known for our passion for – some might say obsession with – talking about the weather. It's a reputation that seems to be backed up by statistics. According to recent research, 94 per cent of respondents admitted to having chatted about the weather in the previous six hours, and just under 40 per cent said they'd discussed it in the past 60 minutes.

Some social scientists say these conversations may be less about the weather and more an equivalent to the physical grooming on which our primate cousins spend so much time. In other words, it's a way of building relationships or initiating contact with others – an icebreaker, if you like. Apparently, there are rules of engagement for these weather conversations: when someone says, 'Raining again?' the expectation is for the person answering to agree. To disagree would be a breach of etiquette. But there again, if you live in Britain – and, in particular, Bristol, as I do – who could possibly disagree with the statement 'Raining again'?

Regardless of why we like talking about the weather, it's something we're all interested in. Just think about all the words and phrases we British use to describe wet weather. The rain could be tipping, bucketing, pelting or chucking it down. It could be a downpour, a deluge, a drizzle, or just spitting. It might be raining cats and dogs, or stair rods. You could say the heavens have opened or it's nice weather for ducks. If you're British, you'll get these references straight away. But even the more cryptic weather phrases are easy for us to understand. As I came back from an overseas shoot once, my taxi driver told me (within a minute or two of my getting in the car), 'The weather hasn't been too clever recently.' Of course, I knew exactly what he meant, but I also remember thinking at the time how confusing that would sound to a tourist coming to the UK for the first time.

Given the amount of rain we have, it's amazing we invented a game that is played outside, sometimes over five days, that depends entirely on dry weather. If you're familiar with cricket you will be just as familiar with the phrase 'rain stopped play'. So, clearly, we get a fair bit of rain in Britain – but then we get a fair bit of everything. Today, as I write this, it's been sunny, windy, rainy, clear and now cloudy. At exactly this time two years ago, my back garden was more than ankle deep in snow. This year we haven't seen so much as a snowflake in Bristol.

Basically (and leaving aside the increasing impact of climate change), the weather in the UK can be summed up in four words: unpredictable but rarely extreme. There is, of course, a reason why Britain's weather is changeable, and it's largely down to our geography – positioned at the edge of the Atlantic and at the point where, due to the Earth's rotation, warm air from the south meets cold air from the north. Our weather is also strongly influenced by the Gulf Stream, a powerful ocean current that starts in the tropical Gulf of Mexico, and which makes Britain's temperature much milder than it should be given its latitude. The warmth of the Gulf Stream also means there is more moisture in the air, and water in the atmosphere makes the weather very changeable.

Above **Grey herons in winter rain at Lake Csaj, in the Kiskunsági National Park, Hungary.**

Britain isn't unique in its changeable weather. Japan, another island nation, also has unpredictable weather (and apparently they talk about it a lot there, too). And it's not just an island thing. I once spent a year making a film in the Badlands of South Dakota – a location close to the centre of the North American continent – and a phrase I heard regularly was, 'If you don't like the weather here, just wait five minutes.' They weren't joking. The town of Spearfish in South Dakota, for instance, holds the world record for the fastest recorded temperature change, minus 20°C to plus 7°C in just two minutes. But, while many of us are obsessed by the weather, what's much more important – particularly to plants and animals – is climate.

WEATHER VERSUS CLIMATE

When people talk about the weather, some might assume they are also referring to the climate, but the two are different sides of the same coin. Weather is what's going on in the skies above one's head, which can change from minute to minute, hour to hour, day to day, whereas climate is what the weather is likely to be over a long period – in other words, the average weather and temperature over time. In a nutshell, climate is what you expect and weather is what you get. Or, to put it yet another way, weather tells you what to wear each day and climate tells you what clothes you have in your wardrobe (sometimes the two converge, like on summer camping trips in Britain where you have to pack for every conceivable kind of weather). Climate obviously varies from place to place – it could be hot, dry, humid, temperate or cold – but it is what shapes animal adaptations, and migratory species depend on it.

Below **A group of male lions wait out a storm in the Maasai-Mara National Reserve, Kenya.**

A BEVY OF BATS

Every year in October, a small patch of forest in Zambia's Kasanka National Park experiences a dramatic transformation. The catalyst is the predictable rains that fall here at this time of the year – every year – kick-starting the ripening of countless fruit trees. It's a food bonanza on which many animals depend. Over the course of four weeks, ten million straw-coloured fruit bats fly in from forests across central Africa, including many pregnant females and those carrying newborn pups. Some of these bats have travelled thousands of kilometres to be here. It's Africa's largest animal migration and the greatest gathering of mammals on Earth.

Below From October to December, a small patch of forest in Zambia's Kasanka National Park becomes home to ten million straw-coloured fruit bats. Shortly before sunset, they head off to feed on the seasonal abundance of fruit – flying up to 65 kilometres per night.

The bats roost in an area no bigger than a dozen football pitches and, as the numbers grow, trees and branches seem to drip with bats. Elsewhere, bats like to space themselves about a wingspan apart while roosting, but not in Kasanka. Here, they are so densely packed that branches often break under their weight – sometimes even whole trees topple over.

At sunset, the bats leave the roost to feed. They fly up to 65 kilometres from the roosts searching for the ripest fruit, with each bat eating nearly twice its own weight every night. Over the course of 12 weeks, they'll consume a staggering 300,000 tonnes of fruit, and while doing so they perform a vital role in regenerating the forests. Indeed, scientists have termed these straw-coloured fruit bats 'the farmers of the tropics', because of their importance in spreading the seeds of many tropical trees.

FALLING TREES

You'd think that filming ten million fruit bats in a small area would be straightforward. Not according to producer Ed Charles, who described the forest where the bats roost as the 'Heart of Darkness'. He said the forest was so dense that you could barely get more than 60 metres into it. And you had to watch your every step as snakes and crocodiles were 'everywhere'. There are leopards too, though Ed and cameraman John Shier didn't see any of these big cats. It was probably because of these predators that the bats seemed nervous of anything moving around under the trees. Certainly, to get any shots of the roosting bats, Ed and John had to approach the trees very carefully and cautiously. The slightest disturbance and the bats would take off and settle elsewhere.

Several times during the shoot, Ed heard the snap of a branch as it gave way under the weight of the roosting bats. Once, a huge tree, with a diameter of over a metre, came crashing down – bats and all. This is clearly why the predators are here in such numbers. But, for the crew, it was also something to be acutely aware of when moving gingerly around the bat-laden trees.

Working under the bats had another obvious downside. Whether roosting or flying, the bats' droppings continuously rained down onto the forest floor – and onto the film crew. There aren't many shoots where the steady flow of faeces and urine from the animals you are filming comes a distant third in things to worry about – falling trees and branches being the first, followed by the snakes and crocodiles.

Fortunately, this wasn't a shoot that required hours in the field every day. After all, there are only so many roosting bat shots one needs in a sequence. The real spectacle occurs 25 minutes before sunset when the bats take off and head out to feed. Then, the sight of countless bats filling the air is enough to take your breath away (and not because of the fine mist of bat pee floating down to the forest floor). For Ed, the big *fly out* was one of the most incredible things he'd ever seen.

Slightly less enjoyable was the accommodation – Zambia's answer to Fawlty Towers, according to Ed. Most guests only stay for a couple of nights – enough to take in the massive bat spectacle – but Ed and John had a three-week stretch. Early in the team's stay, the hotel ran out of vegetables – or, at least, any kind of variety. For some reason, green peppers were never in short supply and, as a result, they came with every meal. Pasta, on the other hand, was a rarity so when 'lasagne' was served, the layers of pasta were replaced by slices of bread. It pays not to be too fussy on shoots in remote locations.

Opposite top **Straw-coloured fruit bats in Kasanka National Park. This is one of the greatest gatherings of mammals on Earth. Here, the bats are being filmed with a drone.**

Opposite bottom **During the day, when roosting, the bats are so densely packed that branches often break under their weight. Sometimes whole trees topple over.**

THE GREAT RAIN MACHINE

Weather may be a catch-all term for the state of the atmosphere in any one place at a given time, but this powerful force also performs a vital role for the planet. Driven by wind, weather distributes freshwater around the globe as rain clouds. Since all life on land is ultimately dependent on freshwater, without this force the planet's plants and animals simply couldn't function.

We live on a largely blue planet but only 2.5 per cent of the world's water is freshwater and, since nearly 99 per cent of that is locked up in glaciers and ice caps or in underground aquifers, that doesn't leave much for rivers and lakes – the water sources on which most of life on land relies. This is where the combination of wind and clouds comes in.

Every second, over 13 million tonnes of water evaporates from our oceans to form clouds, which are made up of trillions of condensed water droplets (a process that starts when water vapour condenses around microscopic particles from volcanic eruptions, dust storms, fires or even pollution, which are floating in the air). This evaporated water eventually returns to the Earth's surface in the form of rain, though – and this is the important bit – in a different place from where it began.

Between evaporation and falling as precipitation, a rain droplet may travel literally thousands of kilometres. This is due to several characteristics of our planet all working together – the shape and tilt of the Earth and its rotation. The spherical shape of the planet, along with its tilt, means that some parts of the planet receive more solar radiation than others. It's why the equator is hotter than the poles. This uneven heating produces air masses of varying temperatures, and winds blow from regions with high pressure (cool, dry air) to regions with low pressure (warm, moist air) in an attempt to even out these differences. If we were living in a windless world, the water droplets would simply fall straight back to the surface. But when combined with the Earth's rotation, these global winds shift rain clouds great distances.

Opposite A Philippine eagle flies over the tropical rainforest at Banaue, Luzon Island, Philippines.

Below In the Tanjung Puting National Park in Central Kalimantan, Indonesia, a female orangutan and her youngster hold leaves over their heads to protect themselves from the rain.

THE GREAT FLOOD

The spin of the Earth and the prevailing winds determine where these clouds reliably blow – meaning some places get less rain and others much more. The Amazon rainforest, for instance, is one of the wettest places on Earth, getting as much as four metres of rain in a year (though a good portion of that is generated by the Amazon forest itself through a process known as transpiration – water vapour produced by the trees). The Amazon River, fed by over 1,100 tributaries, accounts for over 15 per cent of the world's total river discharge into the oceans – more than 14 billion cubic metres a day, all flowing into the Atlantic. The Amazon's flow is so powerful that apparently one could drink freshwater out of the ocean before even sighting the continent.

So much rain falls in the Amazon that every year, starting in November, an area the size of Britain – 250,000 square kilometres – floods, turning the jungle into a seasonal wetland. The inundated areas extend around 20 kilometres from the riverbanks, making this the most extensive system of riverine flooded forest on Earth. The forest can be flooded up to a depth of ten metres, so anything living here must swim or climb.

Where birds once flew, fish now swim between the branches of the forest understory. For vegetarian fish, these are the good times as it's now that many trees fruit, taking advantage of these seasonal residents to spread their seeds. The fish are followed by river dolphins, or botos, which, with their highly mobile necks, are perfectly adapted to hunt amongst the trees and branches.

Even jaguars are found in the flooded forest. These top predators can climb and swim but no one realised until recently that they could live in the flooded forest full time. Researchers have discovered female jaguars feeding, breeding and raising their cubs in the treetops for up to four months, during

Below **Rafting fire ants.** The Amazon's seasonal floods often inundate these ants' underground nests. When this happens, the colony is quickly evacuated, with the workers and soldiers linking together to form a living raft – a process that can be completed in under two minutes. Tapiche River, northern Peru.

the height of the floods. One radio-tracked female was found living in a tree with her six-month-old cub 12 kilometres from the nearest dry land.

As the forest slowly starts to flood, everything living in the leaf litter begins a vertical migration into the trees. But there's one animal that has a different solution to the annual floods. Fire ants live in underground nests but, when their nests become inundated by the flood waters, they do something extraordinary. Workers and soldiers quickly evacuate the larval young and their queen, then start to build a living raft. With the queen safe in the centre, the more buoyant larvae are put in the middle of the raft. These are knitted together by the rest of the colony using a combination of mandibles, claws and legs to link each ant with another. The whole raft, which might be 30 ants deep and contain more than a million individuals, can be completed in under two minutes. Colonial ants, like fire ants, are often described as a single superorganism working together, and this must surely be the ultimate expression of all for one and one for all.

Each ant is covered in fine hairs that trap air against its body. These hairs are not enough to stop a single ant from drowning, but when part of a raft they make the group virtually unsinkable. Scientists have described the effect of the interlinking ants as being like a waterproof fabric, such as Gore-Tex.

The raft can stay floating for days, even weeks, but, being rudderless, it's completely at the mercy of the flood. As they sail through the flooded forest, the ants rotate around the raft like a convection current, presumably to give those at the bottom a break, but researchers have discovered that even those forming the base for long periods survive the experience… or at least don't drown.

Rafts can be attacked by fish. If they split the raft into bite-sized chunks then it could spell the end of the colony. Predatory pond skaters are also a menace, stabbing individuals on the outer edge with needle-like mouthparts, until the ants join forces to drive them off.

Below left **Looking down on a fire ant raft. Each ant links with its neighbour using a combination of mandibles, claws and legs. The queen and the larvae are kept safe in the centre of the raft.**

Below right **New home. When the raft pushes up against suitable vegetation the ants disassemble. They will make this palm their temporary home until the floodwaters recede.**

Opposite **Flooded forest in the Anavilhanas Archipelago on the Rio Negro in the Amazon basin region of Brazil.**

Top right **A predatory pond skater uses its needle-like proboscis to stab a fire ant. Tapiche River, Peru.**

Bottom right **As the raft drifts through the flooded forest, fish attempt to pick off ants and larvae.**

Despite the dangers of both flooding and rafting, fire ants have come to depend on the flood to carry them to new feeding grounds. When the raft is finally pushed against a tree trunk, or other suitable vegetation, the ants disassemble and surge upwards, carrying the larvae and queen with them. They will stay in this temporary home – feeding on other trapped invertebrates – until the flood ends and they can return to the forest floor and dig themselves a new underground nest.

THE SHOOT OF A THOUSAND STINGS

Most people who watch natural history films realise the sequences they've just seen take much longer and are more difficult to film than they might otherwise appear – this knowledge is probably in large part down to the 'making of' programmes that are shown at the end of many of them. However, some sequences still look quite easy. The rafting ant sequence is a case in point. The behaviour is certainly extraordinary and the macro filming is as good as you'll see but, given that the ants instinctively form a raft when flooded, how difficult can filming them really be? The answer, you'll have surmised by now, is very difficult. Indeed, not only was the behaviour tricky to film, it turned out to be one of the most incident-prone shoots we had on the whole series.

Following discussions with a fire ant expert, assistant producer Toby Nowlan decided on a location near Iquitos, in northern Peru. To reach the site

Below **Rafting fire ants.** These rafts can stay floating for days, even weeks, but, being rudderless, they remain at the mercy of the flood. However, fire ants have come to depend on the Amazon's annual flooding to carry them to new feeding areas. Tapiche River, Peru.

took an eight-hour boat journey along one of the Amazon's tributaries. The first job on arrival was to find the right species of ant. This wasn't as easy as expected as there are lots of very similar-looking ants in the area. (There are, in fact, over a thousand species of ant in the Amazon – though nobody knows for sure. There could be many more.)

There is one failsafe way of knowing whether you've found a fire ant colony or not, however, and that is the sudden burning sensation you experience when one bites or stings you – usually both at the same time, though it's the sting that causes the real pain as it's from there that the venom is injected. To state the obvious, they're not called fire ants for nothing. And the burning sting – which sometimes results in pus-filled blisters – was something Toby and cameraman John Brown just had to get used to as the ants boarded both at the slightest opportunity. If the raft so much as brushed against their legs, or the tripod, the ants were off and up. But filming all the details of the raft, like how the tiny ants interlocked with each other, required the crew to

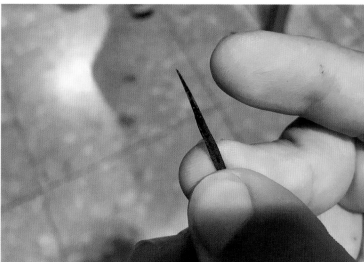

be almost at touching distance from the edge of the disc, which meant there wasn't a lot of room with which to play. And, as the rafts were free floating on the surface, contact with the crew was inevitably fairly regular.

Filming the sequence required a lot of patience – not least because of the steady stream of stings. But, when things went well, it was immensely satisfying. 'As a colony was gradually flooded by the rising waters,' said Toby, 'the ants would pour out of the ground and the surface seemed to boil with the little insects. If we were lucky, when the larvae were carried from their underground nest to a place of safety, the raft would begin to take shape around them and then, as if by magic, the floating disc of ants would suddenly appear before us.'

So far so good, but when the team tried to get shots of the raft from underwater they ran into a rather prickly problem – or one of the team did. The second cameraman, Alex Vale, was wading neck deep in water, trying to get into the right position to film a raft from underneath. Unfortunately, he didn't see the spine-covered palm in the inky water and he walked straight into it. Despite the protection of his wetsuit, he was peppered in spines nearly from top to bottom: 30 in his chest, more than ten each in his hands and wrists and more than a dozen in each leg. Toby and John managed to extract many of the thorns but a good number, particularly those in his wrists, broke off in his skin. Given the concern of infection in the hot and humid environment, it was decided the best thing for Alex was to return to the UK straightaway. Back in Bristol, he went to the rather appropriately named Foreign Bodies Clinic and had his *foreign bodies* from the Amazon surgically extracted.

To get all the shots we needed for the sequence took a month. And after a thousand stings, Toby and John were ready to head for home. As at the start, it was an eight-hour river journey back to the port of Nauta, near the city of Iquitos, in a typical Amazonian riverboat (similar to a British canal boat but much faster), with the 40 cases of filming equipment and other luggage loaded onto the boat's roof. For 7 hours and 59 minutes all went

Above left The crew's capsized boat. Coming into dock, the boat was caught in a combination of strong currents and eddies and, moments later, flipped over. Crew and kit were thrown into the water. Some of the kit was flooded but fortunately everyone survived the experience.

Above One of the spines removed from cameraman Alex Vale, after he inadvertently swam into a thorny palm tree while filming underwater shots of the rafting ants.

well and according to plan but then, just as the boat turned into the dock, it was caught by an unexpected combination of strong currents and eddies. The boat wobbled and lurched, causing the heavy equipment on the top to move in the same direction. It was now suddenly and seriously unbalanced. A moment later the boat capsized and started to quickly fill with water. When this happened, Toby and John were below decks in the cabin and with water pouring in through doors and windows, they had just enough time to grab the small bags containing their passports and swim through the open hatch of the cabin. Along with the boat crew, Toby and John swam to the edge of the river and clung to a bollard. From there, they continued to the jetty, before climbing up a muddy bank to safety. Like most accidents, everything happened in seconds and, though it could have been a lot worse, everyone thankfully escaped unharmed. The same can't be said for the kit. All of it ended up in the river. With help from other boat crews and onlookers, the pelican cases were eventually recovered. Some had got flooded, however, irreparably damaging the equipment inside. Most worrying were the storage drives containing the entire shoot's footage. As a precaution these are always saved onto two drives, which are kept separate from each other. The first set Toby and John found were ruined but fortunately the second set survived the experience. After the shoot of a thousand stings, it would have been a cruel blow to have lost all the footage. There was, of course, a certain irony to this incident. The crew had spent four weeks filming the unsinkable rafting ants, only to end up sinking themselves.

Below **A Peruvian pink-toed tarantula makes for high ground or a tree after the rising river level flooded its silken retreat. Little is known about the species, but it can swim!**

SANDBARS AND GIANT TURTLES

Contrary to popular opinion, it doesn't rain every day in a rainforest. In the Amazon, the very wet season is followed by a less wet – at times, even completely dry – season. The predictability of these weather patterns (or, more accurately, climate) has allowed animals to fine tune their behaviour to very specific periods of the year.

Every October, thousands of female giant river turtles start to gather at locations across the Amazon. With the end of the heavy rains, the river levels

Below **Giant river turtles nesting on sandbars in the Rio Guaporé on the Brazilian–Bolivian border.**

in the Amazon's tributaries start to fall and, as they do so, they gradually reveal sandbars. Providing there's at least a metre's worth of dry sand, these are the perfect nesting sites for the giant river turtles.

The turtles don't lay immediately on arrival. For a week or two, these metre-long reptiles spend part of the day sunbathing at the edge of the sandbar. During this time, their eggs develop inside them. When they're not basking, they're floating at the surface, heads up and watching what's going on around them. As the days go by, the number of turtles continues to build. The biggest nesting beaches attract as many as 50,000 turtles – a quarter of the total population of giant river turtles, though that number is still a fraction

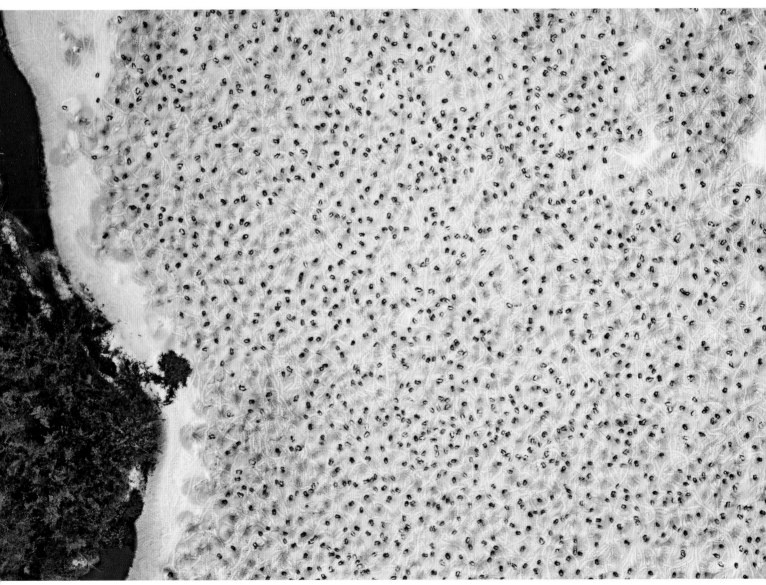

Opposite above left **Prior to nesting, giant river turtles spend much of the day sunbathing on the sandbanks. This is thought to speed up the production and development of the eggs inside them.**

Opposite above right **A nesting giant river turtle. Most of the nesting takes place at night but, here, there are so many turtles that the activity goes on for the first hour or two after sunrise.**

Above left **After 55 days of incubation, a newly hatched giant river turtle emerges above ground.**

Above right **Changes in the Amazon's seasonal cycle of wet and dry mean nests are often flooded before the eggs have hatched. These youngsters are the lucky ones – emerging just before heavy rains flood any remaining unhatched clutches.**

Opposite **Every October, around 50,000 giant river turtles – a quarter of the total population – nest on seasonal sandbars in the Rio Guaporé.**

of what they once were. In the nineteenth century, there were estimated to be tens of millions of giant river turtles before widespread hunting for eggs and meat took its toll.

Egg laying takes place either in the middle of the night or in the early morning, before the sun gets too hot. Once it kicks off, the beaches quickly become carpeted with nesting turtles – so many, in fact, that one wave of turtles inadvertently digs up some of the eggs of previous nesters. They won't go to waste as turkey vultures and caracaras are on standby, ready to take advantage of the sudden glut. Each turtle will lay around a hundred eggs. The whole process of digging and laying will take about an hour, during which she'll be sprayed by sand from other nesting turtles and suffer the irritation of bees drinking salts from her eyes. Once the nest hole has been refilled and patted down, she'll drag herself back to the water and away. She won't return to this spot until the next dry season.

Other turtles will still be waiting for their time to nest. These late arrivals might benefit from not having their eggs dug up by mistake, but if the river rises early the beaches could become flooded before the eggs have completed their 55-day incubation. Giant river turtles are hard wired to begin breeding as soon as their nesting beaches have appeared. But they must leave the rest of the process to fate. Indeed, the future of the giant river turtle is entirely dependent on the reliability of the dry season. Unfortunately, that doesn't always pan out as expected.

FLOODED BEACHES

In 1999, I was on a shoot to film giant river turtles nesting on sandbars in the Xingu River – a mighty tributary of the Brazilian Amazon. It's so large that where the sandbars appeared the river was 14 kilometres wide. You could look out from the beach and barely see land on either side. Over the breeding period, we estimated there to be around five thousand turtles, but that year it was likely that not a single clutch survived. Heavy rains in the headlands caused a sudden rise in the river level, with disastrous results. Most of the nesting beaches were flooded while we were filming, and the turtles that hadn't yet laid had no choice but to abandon nesting for that year.

Fast forward to 2019, the year we filmed nesting giant river turtles for *A Perfect Planet*, and things looked much more promising – though it didn't start off like that. Our first filming location was in the Rio Trombetas, a site that, for 30 years, had some of the biggest gatherings of nesting turtles in Brazil. It was so reliable that the world's foremost authority on these turtles had no hesitation in recommending the location as the top spot. Unfortunately, our experience turned out to be very different.

In 2018, when our first shoot took place, the Trombetas' promised thousands of turtles had crashed to little more than five hundred, and those that showed up were very skittish – a reaction to hunting and other disturbance. We came back with a nesting sequence, though certainly not a good one. Fortunately, further research revealed another, significantly bigger and better site, though much more remote, on the Brazilian–Bolivian border: the Rio Guaporé was said to have as many as 50,000 nesting turtles.

It was a five-day journey from the UK, including an 11-hour drive through '40 years of deforestation', as cameraman Matt Aeberhard put it. He and assistant producer Darren Williams arrived at the Rio Guaporé just as the nesting was getting into full swing. This time the numbers were just as

Below **Cameraman Tom Rowland filming a nesting giant river turtle on the Rio Guaporé.**

predicted. On top of that, the turtles showed no fear of the crew – a sure sign that they weren't being hunted. Indeed, it was very apparent to Darren and Matt that the huge number of turtles nesting on the Guaporé was all down to the careful protection the beaches got from a local community – something that certainly wasn't the case on the Trombetas.

Within a few days of their arrival, there were over a thousand turtles nesting every day – either between midnight and 2 a.m. or from 8 a.m. to 9 a.m. – and, when they weren't nesting, the turtles were basking on the sand or floating in the river. Whenever there was activity on the sandbars, the air was filled with a thudding sound as hundreds of turtles tamped down the sand over their clutch with their bellies, which Darren described as the rather satisfying sound of a job well done.

Despite the huge numbers of turtles and the remoteness of the Guaporé, the Amazon's biggest conservation issue was never far away. Matt and Darren could smell and see smoke every day – the result of the forest burning. It produced something of a paradox for the filming. The haze from the fires veiled the sun, creating a soft, filmic light. Cameraman Matt couldn't help being pleased with this aesthetic but, as he said, 'There's something shocking about this story hidden in the pictures.'

Once again, the dry season ended too early and many of the clutches from late nesters were sadly drowned by the rising river. Early floods like this used to happen once every 20 years. Now, thanks to climate change and a shift in the wet–dry cycle, the turtles have to deal with them every four years.

WEATHER

A MOST UNUSUAL FROG

Winds may carry huge amounts of rain to the Amazon but they don't provide this service to every corner of the planet. One-third of the Earth's land is defined as desert – that is, places with less than 25 centimetres of rain a year, and where evaporation often greatly exceeds rainfall. Many people think of deserts as being predominantly sandy or dune-covered, but this only accounts for around 10 per cent of these dry and super-dry environments. There are rocky deserts, cactus deserts, even polar deserts – it's a variety that is matched by what creates them in the first instance.

The Sahara is probably the best-known desert in the world and, at 9 million square kilometres, one of the largest. The reason for its existence is down to a powerful global weather system – a kind of atmospheric conveyor belt. It starts when warm, humid air rises over the equator; as it does so, it cools. Cool air can't hold as much moisture as warm air, so this is released as rain, which is why we have rainforests across the equator. As this air cools and rises, getting progressively denser, cooler and drier, it also spreads out north and south from the equator, where it eventually sinks back to earth. The dry air now flows back to the equator to start the process again. This circulation of air is also known as the Hadley Cell, after British meteorologist George Hadley. It's the Hadley Cell that creates the planet's subtropical deserts, like the Sahara.

Deserts also form in the centre of continents, like the Gobi in Mongolia, because clouds blowing from the coast are often stripped of their moisture before they reach these parts. Then, there are the coastal deserts, like the Atacama, caused by cold ocean currents, which dry the air above them, and deserts resulting from mountain rain shadows, like Death Valley in California.

All animals ultimately depend on water for survival and those that live in our planet's deserts are no exception – the only difference is that they must be adapted to dealing with very small quantities of it. One animal you might not expect to find in a place with exceptionally low quantities of water is a frog, but they exist in every desert, with the exception of the polar deserts of Antarctica. One of the strangest (and, if one is being anthropomorphic, the cutest) is the desert rain frog found in the arid parts of southern Africa. Like most small desert dwellers, it operates a strict policy of sun and heat evasion, spending much of its life underground in a sandy burrow, and only emerging at night when the climatic conditions are suitable for foraging.

The little rain frog is the shape and size of a marshmallow. The characteristics that make it a good burrower – the short, stumpy legs, paddle-like feet and spherical body – mean that it can't hop. Instead, it waddles across the sand, while keeping an eye out for its favourite food, termites. These tiny insects are 75 per cent water, which makes them a meal and a drink in one; that's if the frog can actually get one into its mouth. Rain frogs aren't the most agile of predators, so it can take several strikes to bag a termite. When it finally manages it, the rain frog must squash in its eyeballs to push the insect down its throat. There's almost nothing about this frog that isn't appealing.

Opposite top **A rain frog – oddly named given its home in the deserts of Southern Africa. The rain frog is the shape and size of a marshmallow.**

Opposite middle **At night, it emerges from under the sand to search for its favourite food, termites, which are a meal and a drink in one.**

Opposite below **Mating rain frogs. The smaller male glues himself to the back of a female and the two then retreat to her underground burrow, where their eggs will be laid.**

Termites may be an important source of moisture for rain frogs, but when it comes to their total liquid needs rain frogs can't get by on termites alone. While it rarely rains in their coastal desert home, there is one relatively reliable source of moisture here – fog. At night, cool air is blown in from the ocean, forming fog banks that shroud the desert in mist. It condenses on vegetation, and the gathering beads of water drop onto the sand below. It's here that the rain frog tops up its water levels – not by drinking it, but by absorbing the moisture through its skin, perhaps through its distinctive 'belly patch' which

Below **Dunes of Sossusvlei**, meaning 'dead-end marsh', so called because it is a drainage basin without outflows, in the Namib-Naukluft National Park, Namibia.

Overleaf **During droughts, Australia's burrowing frogs can survive dehydration by hiding underground in a cocoon of shed skin.**

remains in contact with the sand when the frog is stationary. Scientists believe this translucent, un-pigmented area of skin, dense with blood vessels, helps the frog absorb moisture from the damp sand, acting like blotting paper. Rain frogs are so dependent on this water that they can only survive in places with at least a hundred days of fog a year.

On the foggiest nights, when many other rain frogs emerge onto the dunes, females take the opportunity to find a partner. The breeding calls of male rain frogs are most un-froglike – there's not a hint of a croak, chirp,

grunt or *ribbit*. Instead, it advertises its presence with a low, mournful whistle. If a female expresses interest in the smaller male, he'll glue himself to her back. With such small, stumpy legs, it's the only way he can hold on as she waddles across the sand back to her burrowing spot. (Rain frogs do not need open water to breed. In fact, just to complete their very un-froglike behaviour, they can't swim and would actually drown if dropped in water.) Locked in this embrace, the pair will disappear under the sand where they'll lay their eggs – safe from the desiccating heat of the desert sun. The female will remain with the eggs until they become froglets, which they'll do without passing through a tadpole stage.

Some desert frogs take life underground to the extreme. While waiting for the perfect conditions, Australia's water-holding frog can spend many years underground in what's known as aestivation, a dormant state similar to hibernation. These frogs do need standing water to lay their eggs and, since that might not happen very often in these deserts, the frog must be prepared for the long haul. To prevent any loss of moisture, it shrink-wraps itself with several layers of skin – covering its whole body except the nostrils. As its name suggests, it stores water inside its body – specifically its tissues and bladder – which can see it through as much as five years underground. It's an adaptation that is sometimes exploited by indigenous Australians living in the desert outback. By squeezing the frog you can, apparently, get yourself a mouthful of very potable water.

Desert animals have a huge variety of techniques for cracking the water shortages. One species of Tok Tokkie beetle from Namibia's coastal desert gets the moisture it needs by first climbing to the peak of a dune at night. Once in position, it performs a headstand so that, when the morning fog comes rolling in off the coast, the mist condenses on its shiny black back, the drops of water running down into its mouth. In one night, this fog-drinking beetle can increase its bodyweight by a third with water.

Kangaroo rats, from the American deserts, never drink water. They get almost all the liquid they need from their food – essentially dry seeds (the equivalent, perhaps, of trying to get blood out of a stone). And these rodents have other adaptations for preserving water: their kidneys are so efficient that they only produce small quantities of concentrated urine, and cooling their nasal passages makes the warm air leaving their lungs condense before exiting their bodies, so limiting the amount of moisture lost through exhalation.

Panting may be an efficient way for some animals to cool down but it's not a great technique if you live in a place with virtually no water. It's why camels – the most iconic of desert animals – do not pant. And camels have another trick up their sleeves… or humps. Contrary to popular opinion, camels do not store water in their humps. In fact, they can't store water anywhere in their bodies. What they can do is store fat in their humps and convert that to water when needed. It's one of the survival strategies of the wild, two-humped Bactrian camels of the Gobi – the last truly wild camels and one of the rarest large mammals on the planet, with less than a thousand remaining.

THE LAST WILD CAMELS

At 800,000 square kilometres, the Gobi is the fifth biggest desert in the world (when you include the polar deserts of the Arctic and Antarctic). Although it sits in the heart of Asia, straddling southern Mongolia and northern China, the Gobi's desert status is also due to the rain shadow effect of the mighty Himalayas that prevent monsoon winds from reaching this vast, largely rocky desert.

It's considered to be a cold desert – a title it justly deserves given that, in the middle of winter, temperatures in the Gobi can drop to minus 40°C, with wind chill. But these freezing conditions are just half the story since in midsummer the temperatures can soar to 45°C. One of the few species able to survive these extremes all year round, without hibernating, is the wild Bactrian camel, which must surely be a contender for the toughest animal on the planet.

Wild camels are very distant relatives of the much heavier domesticated Bactrian camel, which number over two million and are still used as pack animals by rural Mongolians. (Recent DNA sampling has shown that the two split more than 700,000 years ago.) The wild camel was once widespread across Central Asia but it's now restricted to the Gobi – and it's well adapted

Below **Wild Bactrian camels in the Gobi desert, Mongolia. These camels are some of the toughest mammals on the planet, able to withstand temperatures ranging from minus 40°C to plus 45°C.**

to this desert's harsh climate. These camels can vary their body temperature by as much as six degrees – a change that would kill most other mammals. (Humans by comparison can deal with a temperature variation of just two or three degrees.) They can also withstand huge water loss – up to 40 per cent of their bodyweight – though, when the opportunity arises, the camels can consume up to 57 litres of water in a single sitting. And, perhaps, most impressive of all, they can smell damp ground from over 50 kilometres away – a very useful skill in a place with virtually no water.

In summer, the camels can't stray too far from waterholes, but in winter they can range far and wide – and they do, across literally thousands of square kilometres. They still need water during this time, but with summer waterholes frozen solid, they must depend on another source – snow, which blows in from Siberia. Where it settles is impossible to predict, and it may not last long – not because it melts (it's too cold for that) but through a process known as sublimation. In the dry air, the snow evaporates without going through the liquid phase. So, to get the water they need in winter, the camels must find and eat snow.

In 2003, while working on the original *Planet Earth* series, I did a shoot in the Gobi desert to film this behaviour. It's still in my top three favourite

shoots, and I've done hundreds. We knew it wasn't going to be easy – not just because the place was vast and there weren't many wild camels, but because they were very nervous of people and, with highly acute senses, were known to run from humans as far away as four kilometres. But we had to get to the Gobi first, and that in itself was a challenge.

Ulaanbaatar, Mongolia's capital, is the only capital in the world where the average yearly temperature is zero degrees – and given that it's very hot in the summer that, of course, means very low temperatures in winter. We got the first taste of this coming out of the city's international airport where the dry cold made the moisture in our nostrils freeze on impact. We all laughed nervously, knowing we would be camping in these conditions in a week.

Leaving Ulaanbaatar, it took us four days to drive to the Gobi desert, sleeping in the cosy *gers* (the Mongolian word for what most people might refer to as yurts) on the way, as guests of extremely hospitable nomadic herders. We passed through a couple of small towns en route, stopping for fuel and supplies or to use the public conveniences. These were mostly of the 'long drop' variety, and in winter they are, strangely, a rather fascinating experience. As a result of the freezing conditions, what you see through the hole in the floor is basically a large stalagmite of poo (one can only imagine what this becomes in the heat of summer). Long drops aside, crossing the roadless interior was seeing nature at its most perfect: the scenery was big, beautiful and virtually deserted.

The Great Gobi National Park is not only one of the largest terrestrial reserves on the planet, it's also totally uninhabited. Astonishingly, we were entering a place as large as the Netherlands and we were the only people in it. I remember this initial sense of wilderness with a giddy excitement. It was a feeling that lasted until sunset.

Opposite **Wild Bactrian camel and calf. The Gobi desert is home to the last thousand wild camels on Earth.**

Below **Wild Bactrian camels are long-distance specialists and are able to smell moisture from as much as 30 miles away.**

Opposite top **Cameraman John Shier scanning the horizon for wild camels in the Gobi desert, Mongolia.**

Opposite bottom **Snow on dunes. The Gobi is considered a cold desert. In winter, temperatures can drop to minus 40°C, with snow blowing in from Siberia.**

Overleaf **During the winter, when waterholes are frozen solid, the wild camels depend on windblown snow for their moisture.**

After several nights in the cosy, camel-skin *gers*, the first night of camping was a shock. There was no snow, but the temperature dropped to minus 25°C, and I found it impossible to keep warm despite ridiculous layers of clothing. The nights, I have to admit, were miserable but the days were glorious.

We had our first sighting of camels early on, making me think the challenges had been overestimated. But getting close was a different matter. The camels were spots in my 10x40 binoculars and any attempt to reduce the distance made them run. Then we woke one morning to find a Siberian wind had dumped several inches of snow over the Gobi's distinctive black gravel plains and rolling hills. In one night, the Gobi had been turned into an Arctic wilderness. The transformation was otherworldly, and the fresh snow made it easier to track the camels.

Trying to second-guess the movements of a herd that we'd spotted in the distance, we parked our vehicle at the base of a large hill and climbed halfway up. Ahead, we could see the camels, in a line, padding across a vast snow-covered plain with a backdrop of snow mountains. Camels and snow – the sight was sublimely surreal. Unfortunately, this turned out to be one of just six filming opportunities we had in two months.

Sixteen years on from *Planet Earth*, it seemed the right time to revisit the remarkable wild camels of the Gobi for *A Perfect Planet*. Not only were they a textbook example of a desert-adapted species but new technology, like drones, would be able to give the original sequence a significant reboot. When research on the topic started, the disappointing news was that the number of wild camels hadn't increased since my shoot in 2003. Indeed, for ten years after the filming their numbers had dropped steadily – by an estimated 30 animals a year. However, thankfully, things were beginning to change and with a team of local scientists monitoring them and better levels of protection, the declines have been reversed and the trajectory seems to be now gradually on the up.

Nevertheless, the ratio of camels to area was still mindboggling, so finding them was going to be no easier – clearly a big concern for a wildlife film crew. Some other things hadn't changed either. It still took around five days for the team to reach the Gobi from Ulaanbaatar – though for the *Perfect Planet* team, led by producer Ed Charles, they didn't need the hospitality of local herders for the overnights as they were travelling in large camping trucks, which they would be using throughout the shoot. The crew also stopped at a couple of towns on the way down – the same ones I had stayed in years earlier – and it was nice to hear that the long drop stalagmites were still a feature.

Ed and the team saw nothing for five days, then had brief daily sightings for ten days, then nothing for another five days. Many of the sightings were of camels running away. Hunting pressure remains an issue – so the camels were still very nervous of people. At one point, when the terrain got too rocky, the crew had to abandon their cosy, but less manoeuvrable, camping trucks and resort to tents (and experienced, what it's like to sleep at minus 25°C with only fabric protecting you from the elements). Halfway through the shoot, Ed found himself wondering whether getting the sequence they'd aimed for was actually impossible. But, to use that old *making of* cliché, perseverance eventually paid off!

A CURTAIN OF CRABS

The biggest weather system on the planet is the Asian or Indian monsoon. Monsoons are seasonal winds that always blow from cold to warm regions. Since they appear at the same time every year, they shape the climate of the places in their path.

In April, as the Indian subcontinent heats up, it draws cooler air from the ocean, marking the start of the monsoon. The winds make landfall in southern India at the beginning of June, before moving across the subcontinent and, over four months, dropping huge quantities of rain – water that one and a half billion people and countless animals totally rely on. At the monsoon's peak, 17 million tonnes of rain fall on the subcontinent. In some places, these seasonal rains represent more than 90 per cent of the annual rainfall.

Almost as big is the Australian monsoon, which, apart from bringing rain to its namesake, is also crucial for a tiny island over 1,500 kilometres to the west of Australia. Christmas Island (so called because it was officially discovered on Christmas Day in 1643) covers 135 square kilometres and is home to just over 1,400 people, a number that pales into insignificance when compared to its world-famous inhabitant – the red crab, which has an estimated population of

Below **In good years, countless millions of baby red crabs arrive back on the shores of Christmas Island – so many that they can turn the beaches red.**

Above **Christmas Island's red crabs spend their lives on land – and are unable to swim – but to breed they must return to the sea. Each crab will release around 100,000 eggs into the ocean.**

around 50 million (though a big drop from the 120 million that was calculated in the 1980s and 1990s). All depend on the predictable rains that start in December.

The crabs spend most of the year hanging out in burrows or rock crevices, where they lead a solitary existence. If the humidity levels are high enough, they will come out to feed – a diet that consists largely of fallen leaves, fruit, flowers and seedlings, which are recycled back to the forest floor as fertiliser. If the air gets too dry, the crabs retreat to their burrows and plug up the entrances to maintain the humidity inside. Crabs breathe through gills and these must be kept moist.

As far as land crabs go, this is all *so far, so normal*. What makes Christmas Island's red crabs so remarkable is their annual mass movement to the coast to breed – considered by many (not least Sir David Attenborough) as one of the most spectacular animal migrations on Earth. When the monsoon hits the island in November, the relative humidity is high enough for the crabs to embark on their epic trek to the sea, which for some may be as much as ten kilometres, and they do this in their millions. They must time this journey to coincide with a specific phase of the moon – a receding high tide during the moon's last quarter – as this is the best time for the crabs to release their eggs into the sea. The speed of travel depends on this lunar timetable. If time is on their side they might stop to eat and drink on the way. If not they'll gun it to the coast, with surprisingly little meandering. Migrating crabs fitted with radio

transmitters have been tracked walking a direct line for over four kilometres, even over very rough and undulating terrain.

And they don't just head to their nearest stretch of coastline. Research has also shown that crabs seem loyal to a particular location. Why, nobody knows for sure, but it could be because that is where they came ashore as tiny crablets (or *megalopae*, as they are known scientifically). Occasionally, they run the gauntlet of their much larger cousins, the robber crabs, whose powerful pincers can make short work of a red crab. If, during their trek, the weather gets too dry then the crabs will hunker down in the shade to avoid dehydration.

The males are the first to arrive at the coast and, after a dip in the ocean to dampen their gills and replenish their water supplies, they retreat to the sea cliffs where they will dig a burrow and wait for the females. It's in these burrows that mating will take place and where, afterwards, the females will hang out until they're ready to spawn, around two weeks later.

When the right tide comes around, the females emerge from their temporary burrows in a mass synchronous movement, described by David Attenborough as 'a great scarlet curtain moving down the cliffs and rocks

Above **Every year, millions of Christmas Island crabs migrate to the coast to breed. Here, hundreds are descending a cliff face on their way to release their eggs into the sea.**

towards the sea'. Since they can't swim, they approach the water with understandable caution but fear is soon replaced by a gay abandon as each female shakes 100,000 eggs into the tidal flow. Some wedge themselves into the sand, others cling tight to rocks and sea walls. But whichever spot they choose, the spawning dance is the same – a comical and rhythmic jerking of the body while front claws are held aloft. Those that lose their grip may be washed away and drown but, for the vast majority, it'll be back to the home burrow after their eggs have been released.

As soon as the eggs hit the seawater they hatch into free-swimming larvae and drift offshore – many into the waiting mouths of predatory fish, like manta rays and whale sharks. This, however, isn't an issue. The crabs' synchronised spawning means that there are so many larvae that it would be impossible for even these large filter feeders to hoover up all of them. After four weeks out at sea, the larvae return to begin a life on land. Nobody knows why, but every ten years there is a truly vast return of *megalopae*, turning long stretches of the coast red. Then, in tightly packed masses, the hordes of juvenile crabs sweep slowly inland.

ROAD CLOSED, CRABS ON THE MOVE

Christmas Day on Christmas Island – for the shoot's field director, Amy Thompson, that was a rather appealing notion. After all, there can't be many people able to have that on their tick list. As it turned out, it wasn't to be – not quite, anyway. Crab activity meant Amy and her camera team – Sophie Darlington, Ryan Atkinson and Braydon Moloney – actually flew out of the island on Christmas Eve. But while they weren't able to claim the ultimate Christmas Day retreat, they did manage to tick off pretty much everything on the shot list – and that is almost as rare as spending Christmas Day on Christmas Island.

When the red crab migration begins, it becomes the main topic of conversation on the island – with the locals talking about the crabs like we in Britain chat about the weather. This annual event is something the Christmas Islanders are very proud of – though, unfortunately and despite best efforts, hundreds of thousands of journeying crabs are still killed on the roads each year.

In the 1980s, it was estimated that as many as a million migrating crabs were killed on the roads each year. To address this problem, fences, bridges and under-road crossings were built and these, with the addition of road closures at peak times, have significantly reduced the number of crab deaths – though it still runs into the hundreds of thousands. Much more challenging has been the mortality caused by the so-called yellow crazy ants – an invasive species that has caused mayhem to the island's ecology. The tiny ants can kill the crabs with formic acid and overwhelm them by sheer numbers – just one hectare of forest might contain 20 million yellow crazy ants. It's estimated that since their accidental introduction, the ants have killed as many as 15 million red crabs – more than a quarter of the total population. But scientists are now fighting back and hoping that a strategy of biological control will keep the ants in check.

While road closures across the island were great for the migrating crabs, they weren't quite so good for a film crew with lots of equipment. However, since crabs are only active at certain times of the day (early morning and late afternoon, when it's not too hot to migrate), outside of those times the roads remained open. This meant that the crew could get into position, with all the kit, ahead of any road closures at 3 p.m. This was particularly important for the spawning, when peak activity occurred from late afternoon to sunset and then from 2 to 4 a.m. By that time, the roads would be closed, which meant the team had to sleep in the car. At first, they thought it would be for just one night – two at the most. But then two became five.

It was decided that Amy and Sophie would sleep in the front, Ryan and Brandon in the back. To create a more comfortable sleeping temperature they slept with the doors open on the first night and everyone drifted off to the sound of thousands of crabs padding gently across the leaf litter (not dissimilar to the pitter-patter sound of rain, according to Amy). But, in the middle of the night, Sophie was woken by a large robber crab climbing into

Above **A road is closed to traffic on Christmas Island, as it's on the red crabs' annual migration route between the forest and the sea.**

Opposite **Cameraman Ryan Atkinson filming red crabs on the move. The annual migration of Christmas Island's crabs is one of the greatest natural history spectacles on Earth, with tens of millions making the journey to the coast.**

the footwell by her feet. The pincers on these crabs are strong enough to sever fingers so on the second night stuffiness gave way to security.

With crabs everywhere, close interactions were inevitable. During the spawning, crabs clung to the tripod, light stands and legs of the crew – convenient places to spawn from while in the surf but, given the crabs' inability to swim, they were probably hanging on for dear life too, knowing that one rogue wave would send them to the bottom of the sea. (In fact, while filming the spawning, a large and unexpected wave took out one of our cameras.) It wasn't just over the spawning that the team got up close and personal with the crabs. If you sat in the leaf litter, crabs moved over you; if you left a bag on the ground, crabs crawled in. At the end of the shoot, and before getting on the plane for home, the crew gave all the cases a very careful search for accidental stowaways. Very occasionally foreign critters, particularly small ones, appear out of cases in our camera department. A red crab appearing out of a case on Christmas Eve would certainly be a surprising Secret Santa gift for the person opening the lid…

BEE-EATER TOWERS

For a little over 10,000 years (a geological era known as the Holocene), life has adapted to the predictable cycles of weather, or climate. Much of Africa, for instance, is driven by the rhythm of a wet and dry season – particularly the southern and eastern parts. Nowhere is this more obvious than at Victoria Falls, on the border of Zambia and Zimbabwe, one of the world's greatest waterfalls. Here the mighty Zambezi River flows over a drop of just over 100 metres. At the peak of the wet season, in December, 5,000 tonnes of water cascade over Victoria Falls every second. The spray can rise to a height of 400 metres and is visible up to 30 miles away. There is so much mist in the air that the view of the Falls is often obscured, making sense of the Falls' African name, *Mosi-oa-Tunya*, which means 'the smoke that thunders'. Five months later, however, and the thunder of water literally turns to a trickle. It's almost as if someone has turned off the tap. But, this massive change is entirely natural. It's part of a reliable cycle of wet and dry that everything living in this region of Africa is adapted to – with winners and losers during both periods.

As the dry season takes hold, the Luangwa River, a major tributary of the Zambezi, dwindles into pools. Riverbanks that were once underwater are now exposed, and that is the cue for carmine bee-eaters, which depend on these sandbanks for breeding, having flown here from the Congo jungle hundreds of miles away. The banks are the perfect location for their nest burrows – out of reach of any predators above and too steep for anything to

Above **A family of elephants crosses the Luangwa River, during the dry season in Zambia.**

Opposite **The Zambezi River, Zambia. At the height of the wet season, in December, around 5,000 tonnes of water a second cascade over Victoria Falls. By May, during the dry season, this flow turns to a trickle.**

Opposite top **A flock of southern carmine bee-eaters flies from a river cliff in the South Luangwa National Park, Zambia. This is the biggest colony of southern carmine bee-eaters in the world.**

Opposite below **During the dry seasons, carmine bee-eaters nest in the Luangwa's exposed sandbanks. The bee-eaters attract fish eagles, whose strategy is to pin their prey against the riverbank.**

Below **A pair of carmine bee-eaters look out from the burrow they have dug to raise their chicks.**

Overleaf **A group of puku approach the edge of the riverbank where carmine bee-eaters are nesting.**

climb up from below. And even if they could, the would-be nest raider would have to cross the shallow river at the base of the banks, which is patrolled by large crocodiles. The dry season is a good time for these massive reptiles too. Animals are drawn to the shrinking river to drink and the crocs know it, concealing themselves in the shallow water until they're ready to strike. Even a flock of thirsty Quelea – a sparrow-sized bird that gathers in huge numbers at this time of the year – is worth leaping for. As is a bee-eater trying to get a mouthful of water while on the wing or fighting with a nesting neighbour and inadvertently tumbling into the river.

Bee-eater burrows can extend for two metres into the soft river sand and in sought-after sections of bank there can be as many as 6,000 nests with barely a bee-eater's wingspan between the holes. Unfortunately for the bee-eaters, their annual congregations attract other agile predators – African fish eagles. These birds of prey don't just eat fish. They're rather partial to bee-eaters too, particularly at this time of the year, when they also have young to feed. An eagle fly-by along the bank sends shockwaves of panic through the colony. When an eagle is on the hunt, the smart thing for a bee-eater to do would be to retreat into their burrows but, instead, most of the colourful birds flee in disarray. The eagle targets any that are too slow to bolt, pinning them against the riverbank.

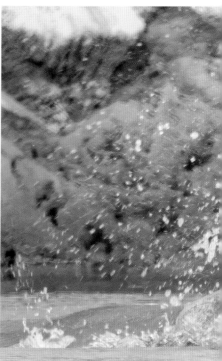

JUMPING CROCS AND COLLAPSING BANKS

Hunting behaviour is often the trickiest behaviour to film – not least because when a predator goes into hunting mode, it can happen without notice and be over very quickly. Sometimes the only strategy is to find a good position, stick with it and watch continuously.

To film crocs hunting, cameraman Tom Rowland spent every day of the shoot in a boat hide on the river in front of the colony. To minimise any disturbance to the carmines, his local assistants paddled him to the hide in a canoe before sunrise. During the five-minute canoe ride, Tom's head torch invariably picked up the eye shine of hippos and crocs in the water all around. It was something Tom tried not to think about too much while sitting on his own in the hide during the day – particularly when, on occasion, one or the other would come to within a couple of metres of him.

On his very first day in the hide, Tom saw a crocodile jumping for a bee-eater, leading him to think the behaviour would be easy to film. The reality check was not long in coming. Nothing happened on the second day, nothing again on the third day – nothing, in fact, for the following two weeks. The crocodiles drifted around occasionally, but showed little interest in leaping for food (at least during the day, when Tom was watching).

Fortunately, there were the nesting bee-eaters to film so, when the crocs were basking on the sand and showing no interest in hunting, Tom focused on the colony. At one point, he heard the sound of what he thought was a

Above left **A section of the riverbank on the Luangwa River collapses – taking with it hundreds of carmine bee-eater nests.**

hippo running into the water. Swinging around in his hide to see where the noise came from, he saw the tail end of what had caused it – part of the bank falling into the river. He watched the area for a while longer but nothing else happened.

A couple of days later, by pure luck, he was looking at the nesting colony and he saw a tiny piece of bank falling into the water. His interest was roused so he framed up on a large part of the colony and put the camera in pre-roll (a function that allows the camera to continuously record for several seconds at a time, so that you don't miss the beginning of something very unpredictable; at the same time, you also don't end up with lots of footage where nothing is happening). Five minutes later, a large section of the bee-eater bank collapsed into the river, taking with it hundreds of nests. A second after it started, Tom pressed record and, as a result of the pre-roll, got the whole event from the start. It was a very dramatic shot and a much bigger collapse than any of his local field assistants had ever seen. Perhaps it would make up for the lack of success with the jumping crocs, Tom thought, as the shoot was rapidly coming to an end. But then, as Tom said, 'In the last hour of the last day, with the last battery on the camera, in the last bit of light, the crocodile I'd been watching carefully for three hours decided to jump five times for bee-eaters.' The crocodile wasn't successful but the jumps were still thrilling. And in the world of wildlife filming, a few minutes of success is all it takes sometimes.

WILD, WILD WEATHER

Most of us are aware that the planet's weather is getting worse, but that has more to do with the frequency of extreme events than their severity. The Holocene (the geological era encompassing the last 10,000 years) has been characterised by reliable weather patterns, or climate. This predictability has allowed humans the luxury of planning (having, for instance, specific months when seeds need to be sown and crops harvested) and it is the principal reason why human society and culture have been able to grow so rapidly in a relatively short period of time. (By comparison, in the 150,000 years before that, humans barely advanced their culture at all.) Nevertheless, planet Earth has always been subject to extreme weather, such as hurricanes, tornadoes, lightning, dust storms, floods and drought.

Hurricanes, cyclones and typhoons are different names for large tropical storms. What you call them depends on where you live: the storms that are generated in the Atlantic and north east Pacific are hurricanes; those in the Indian Ocean are cyclones, and typhoons occur in the Western Pacific. The ingredients for a hurricane include warm moist air from tropical waters, low-level winds and enough distance from the equator to cause a spin. (The Coriolis forces which generate the spin do not occur on the equator.) It's the spin – which in the Northern Hemisphere goes counter clockwise and vice versa in the Southern Hemisphere – that creates the eye of the storm, a region of mostly calm weather that could be 30 to 65 kilometres in diameter. In a hurricane, the most severe weather with the highest winds occurs on the wall of the eye. In a category 5 hurricane (the highest category) wind speeds can be greater than 250 kilometres per hour and can expend as much energy as 10,000 nuclear bombs. These megastorms can destroy houses and make entire areas uninhabitable for weeks or months. The largest hurricane ever to hit the United States was Hurricane Sandy in 2012, which was 1,500 kilometres wide when it slammed into New York and New Jersey.

While a hurricane starts life out to sea, a tornado forms over land as part of a supercell thunderstorm (you can have a supercell that doesn't create a tornado, but not the other way around). Both hurricanes and tornadoes are associated with violent, rotating columns of air, but there are key differences between the two: tornadoes are much smaller than hurricanes – most being between 20 and 100 metres wide, though some of these swirling vortexes can be over 3 kilometres. Also, tornadoes generally only last a few minutes and travel an average of just a kilometre or two. The United States has the dubious honour of having the most tornadoes of anywhere in the world – around a thousand a year – and most of those occur in the Great Plains (or, more colloquially, Tornado Alley). The most deadly tornado ever to hit the USA was the 'Tri-State Tornado' in 1925, which travelled for over 350 kilometres, making it the longest ever recorded, and killed 695 people. The fastest ever recorded wind speed from a tornado was 484 kilometres per hour, measured in May 1999.

Opposite **An F-4 category tornado, near Campbell, Minnesota, has the lower portion of the funnel illuminated by the sun.**

Across the world, at any given moment, it's estimated that there are around 2,000 thunderstorms on the go – that's 16 million a year. The most thundery place on the planet is Java, in Indonesia, where storms occur on average 220 days per year. With a thunderstorm comes lightning, a bolt of which can travel at speeds of up to 218,000 kilometres per hour and reach temperatures of up to 30,000°C – hotter than the surface of the sun. The lightning capital of the world is Lake Maracaibo in Venezuela, which gets 300 nights of lightning a year.

And then there are the dust and snow storms, the floods and droughts. The planet's dust storms produce as much as 200 million tonnes per year, with the biggest able to travel for thousands of kilometres across oceans. A single snowstorm can drop as much as 40 million tonnes of snow. (The largest hailstone dropped in South Dakota, USA; it measured 20 centimetres and weighed nearly a kilo.) The longest recorded drought, on the other hand, took place in the Atacama desert in Chile, where for 400 years – between 1571 and 1971 – no rain fell. Lack of rainfall is also thought to have contributed to the demise of the Egyptian empire. Too much rain can cause even greater problems. The deadliest flood in the history of humanity occurred in China in 1931, and was estimated to have killed up to four million people.

But as devastating as these weather events can be, they are also part of a natural system. Hurricanes, for example, help to maintain the balance of global heat by redistributing warm tropical air from the equator to the poles and are important rainmakers, irrigating land and recharging aquifers. Dust is rich in nutrients and minerals like nitrogen and phosphates, and wind-blown dust storms distribute them around the world. Indeed, the Amazon's fertility is partly down to Saharan dust blowing across the Atlantic – as much as 22,000 tonnes a year. Lightning fixes nitrogen into nitrates, which plants depend on to grow. Lightning strikes are also a crucial generator of ozone, which protects us from the harmful rays of the sun.

Overleaf **An ominous bank of mammatus cloud is downwind of a super-cell thunderstorm above the Standing Rock Indian Reservation, near Little Eagle, South Dakota.**

Right **Lightning from a passing electrical storm lights up spectacular rock formations at Badlands National Park, South Dakota.**

Above **A herd of African savannah elephants walks towards a storm in the Maasai-Mara National Reserve, Kenya.**

FORECASTING THE WEATHER

Knowing what the upcoming weather is going to be like is not just useful for people. For animals, predicting a major storm, for example, could mean the difference between life and death, but can they forecast the weather? There's plenty of folklore and anecdotal information pointing to the positive. Some Native Americans, for instance, believe a hare's feet will grow fluffier if there are heavy snows in the offing and that black bears decide on where they sleep in a cave depending on how cold the winter will be. And, of course, there's the old wives' tale about cows lying down when it's about to rain. But the science backs up some of these tales.

Nobody assumes that animals have a sixth sense that can predict weather, but there's plenty of evidence to suggest that they're able to make better use of the five senses. The most critical, perhaps, is hearing. Some animals can hear sounds at either end of the spectrum that can't be detected by humans, like infrasonic sound below 20 hertz. Scientists believe that hearing in this

infrasonic range is how animals seem to detect changes in the weather. It's believed, for instance, that elephants are capable of hearing the infrasound of thunderstorms – a useful skill for locating ephemeral sources of freshwater, and particularly important for desert elephants. Sharks also seem to be able to sense an approaching storm or cyclone. Studies on tagged sharks have noted evasive behaviour before a storm, such as juvenile sharks leaving shallow water nurseries to temporarily flee to deeper waters several hours ahead of any wild weather. This ability is likely due to the system of sense organs found along a shark's lateral line, which is used to detect the movements of prey but which may also sense changes in air pressure.

Migrating birds also seem to be able to divert their course way ahead of any approaching storms. And it seems there is scientific evidence proving that cows really do lie down in response to the onset of rain. On hot days, cows stand up to allow more of their surface area to lose heat to the air but, conversely, lie down to retain heat during cold spells – and, since cooler temperatures precede rain, it makes sense that when cows are lying down, it's an indicator of oncoming rain. And it's not just animals – some plants and fungi can also forecast wet and dry weather. Dandelions and clover, for instance, fold their petals ahead of rain and some mushrooms expand prior to rain.

With major – and potentially destructive – weather events becoming more frequent, it's even more important for people to be able to accurately forecast future conditions. Until relatively recently, predicting the weather even a day or two ahead seemed rather hit and miss – unless you were living in a place like Spain's Andalusia, for example, which gets 300 days of sunshine a year. In many countries, forecasting something like rain is given as a percentage, which, if one was being cynical, means the forecasters can never be wrong. Nevertheless, a five-day forecast today is apparently as accurate as a one-day forecast 40 years ago. But things are set to get even better.

The British Met Office is investing in a billion-pound supercomputer – allowing billions more daily observations from satellites, weather stations and ocean buoys (even now the Met Office's forecasting is based on 200 billion data points a day). Currently, the Met Office's simulated forecast is based on data coming from grid points that are ten kilometres across. When the new supercomputer is up and running, the aim is to have a finely detailed model of everything from the winds to temperatures and pressures, based on grid points of just 100 metres across. Then, planning a weekend BBQ or picnic in the countryside will, at last, be straightforward.

Opposite **At dusk, blacktip reef sharks congregate in shallow waters around Aldabra Atoll, in the Indian Ocean.**

Below **NASA's Skylab Orbital Workshop is seen from the command and service module above a pale blue Earth.**

OCEANS

LOST AT SEA

The life of a rubber duck – that most iconic of bath toys – is a limited one, its outlook seldom extending beyond the steep white sides of a bathtub. In January 1992, however, 28,000 small rubber ducks achieved an unexpected broadening of their horizons, when the shipping container in which they were packed fell overboard on a journey from Hong Kong to the USA. Written off as a commercial loss, nobody gave them another thought. They were gone, lost at sea. But, as it turned out, that was just the beginning of their story – one that was to shed new light on the powerful systems that drive our oceans.

As anyone who has ever pushed one around a bath knows, rubber ducks are very good at floating and, once the shipping containers broke up in the Pacific swell, this is precisely what the 28,000 did with their newfound freedom out on the open ocean. But even that might have been the end of the tale were it not for another standard feature of rubber ducks: they're virtually indestructible, so they have a long life expectancy. Over the next 15 to 20 years, the ducks travelled halfway around the world, their progress being charted by scientists and beachcombing members of the public. Indeed, since their release, it's been estimated that they've floated over 27,000 kilometres, pitching up on the shores of the Pacific Northwest, Hawaii, Alaska, South America, Indonesia and Australia. They were even found in ice in the Arctic, having floated through the Bering Strait. From there, some popped up in the Atlantic, eventually making it to beaches in Scotland. So, from a hub in the middle of the Pacific Ocean, the adventurous bath toys have landed on five different continents – and therein lies the real point of this story.

The incredible journeys made by the rubber ducks have helped oceanographers to have a much better understanding of different ocean currents. Knowing where the ducks started their journey, and plotting their progress over the following two decades, scientists have been able to determine how big some of these ocean currents are and how long they take to complete a circuit. The information has helped create computer models that can predict, amongst other things, the distribution of fish and plankton, or, perhaps, the likely flow of a large oil spill.

The *Friendly Floatees*, as researchers named the ducks, have also provided vital information on ocean gyres, which are large circulating currents: in particular, the North Pacific Gyre that stretches from Japan to Alaska, where it's thought a couple of thousand rubber ducks are still bobbing around on the waves. Thanks to these ducks, researchers now know that it takes around three years for the currents in the gyre to make a full circuit. The work also threw a spotlight onto one of the great problems facing our oceans today – plastic litter. Vast quantities of plastic have been caught up in these circulating currents, and, as a result, the North Pacific Gyre has become better known as the North Pacific Garbage Patch or Trash Vortex.

Opposite **Off the coast of Portugal, an ocean sunfish eats a plastic bag suspended in the water, mistaking it for its normal prey of jellyfish.**

MOTION IN THE OCEAN

The story of the rubber ducks also reveals an important truth about the planet's oceans – there are not really five oceans, but one. Stick your toe in, say, the cool seawater off Cornwall and you're instantly part of a single global system. The waters of this one ocean are constantly on the move through a huge network of currents, as the journeys of the *Friendly Floatees* proved.

Flowing like rivers, currents move in all directions across the planet and are driven by a variety of forces: wind, differences in water density and tides. Some of these currents are cold, while others are warm; some move short distances, others for tens of thousands of kilometres; some are narrow, others hundreds of miles wide. And many vary in strength depending on the time of the year. But, together, these currents are vital for the health not only of the ocean but of the whole planet, affecting both the distribution of nutrients in the sea and the world's climate. The Gulf Stream, for instance, is a warm ocean current, driven by wind, that flows from the Gulf of Mexico, up the eastern seaboard of the USA and then across the Atlantic to Northern Europe. It is one of the strongest current systems in the world, transporting around thirty times the amount of water of all the rivers on Earth. It circulates warm water from the subtropics to Northern Europe. Without it, the British Isles would have significantly colder winters – in fact, they would be more like those in north-eastern Canada, which sits on roughly the same latitude as Britain.

Below Swirls and eddies catch the sun's rays on the surface of Canada's Gulf of St Lawrence, as seen from the International Space Station.

Right **Starting at the poles, the global conveyor belt circulates water around the planet. Every drop of seawater rides these currents, taking a thousand years to complete a single circuit.**

The most important kinds of ocean current are those that move over the surface and those that flow along the deep. Surface currents are largely powered by the wind – the direction of which is mostly determined by the planet's rotation, or the Coriolis Effect. As a result of this force, winds and currents are deflected: in the Northern Hemisphere, it causes them to veer to the right, while in the Southern Hemisphere they curve to the left. Deepwater currents, on the other hand, are driven by variations in seawater density, starting mostly at the poles.

The most important movement of water on Earth is a deepwater current known as the global conveyor belt – though referred to by scientists as a thermohaline circulation (*thermo* meaning temperature, *haline* relating to salinity). The conveyor starts in the poles, where the freezing conditions result in cold, salt-rich water sinking into the deep. This happens because salt is mostly expelled when sea ice forms, so the water left behind is both saltier and more dense. As less dense water moves in to replace this heavier salty water, it sets in motion a current that literally circles the planet's oceans, moving between the deep and the surface and transporting an immense volume of water – more than a hundred times the flow of the Amazon River. It's said that every drop of seawater on Earth rides this conveyor, though very slowly. Moving at a rate of a few centimetres a second, scientists estimate that it takes as much as a thousand years for the current to complete a full circuit. On its journey it transports heat, nutrients and dissolved gases, like carbon dioxide, making this massive flow of water one of the planet's most important life-support systems.

THE DRIFTERS

Opposite **A green sea turtle hatchling leaves its nest site on Karan Island and heads for deeper water in the Arabian Gulf.**

Below **A release of European eels heads downstream during a scientific study of fish stocks at Port de Sète, France.**

Anyone who has spent time sailing in the open ocean – or been unlucky enough to have found themselves left at the mercy of wind and currents following a shipwreck – knows that most of this habitat is a vast blue desert. You can sail through hundreds of miles of open ocean and barely see a living thing. Life depends on nutrients, and the availability of these is largely determined by the flow of currents. Put simply, where you have currents you have life.

Many smaller marine creatures spend all or part of their lives drifting on ocean currents. In Pixar's *Finding Nemo*, the namesake of the film – a reef clownfish – encounters the East Australian Current (EAC), where he joins a group of turtles surfing through the ocean on a warp-like flow. Unsurprisingly, a fair degree of Hollywood creative licence has been employed in this scene, but the basics are actually correct. It may not be a narrow jet, like in the film, but the EAC is one of the strongest currents in the South Pacific, moving at speeds of up to 7 kilometres per hour. It transports 40 million cubic metres of water every second in a flow that's almost 100 kilometres wide. And currents like these do carry a lot of ocean creatures.

After swimming offshore for a day or two, hatchling turtles, for example, are thought to spend the first stage of their lives going with whatever flow they encounter. European eels do much the same thing, though there's more purpose to their journeys. The eels are born in the Sargasso Sea, out in the middle of the Atlantic Ocean (an area also known as the Bermuda Triangle), and soon after hatching they hitch a ride on the Gulf Stream. The larval eels

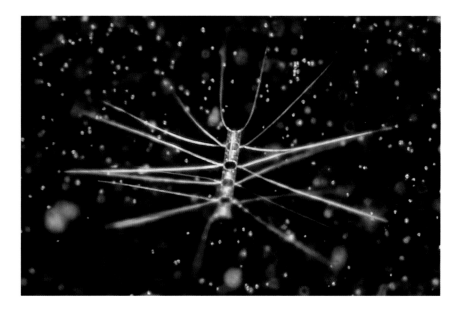

drift for thousands of miles on this huge current all the way to the UK and mainland Europe, before heading up rivers where they will spend most of their adult lives. They are so dependent on the Gulf Stream and other surface currents that the smallest changes in the direction and speed of these currents can have a significant, and often negative, impact on eel numbers – indeed, the normal peaks and troughs in eel populations seem to have been replaced by a steady decline, which is thought to be due to long-term changes to the Gulf Stream as a result of climate change.

The most important ocean drifters of all are plankton (the name derives from the Greek word *planktos*, meaning wandering). The most numerous are phytoplankton, like diatoms and dinoflagellates – microscopic marine algae and bacteria that are found at the sunlit surface. Like plants on land, they contain chlorophyll to harness sunlight, using photosynthesis to turn it into energy. There are over 4,000 species of phytoplankton and together they produce more oxygen than all of our forests combined – making them a crucial ally in our fight against climate change.

Dinoflagellates use whip-like tails to help them move through the water, but their tiny size means they are still at the mercy of the currents they use to travel through the oceans. And currents are vital to phytoplankton for another reason. In addition to sunlight, their survival also depends on nutrients brought up from the seafloor by ocean currents. When conditions are perfect phytoplankton can multiply in astonishing numbers, turning the ocean green. When they die, they sink to the seafloor, joining countless other particles of organic matter – known collectively as marine snow. It can take up to 30 years for some of these particles to reach the deepest parts of our oceans, where together they form a rich bed of sediment: the basis of the ocean's nutrient cycle. And it's cold, upwelling currents that bring these nutrients back to the surface, where they feed fresh blooms of plankton, starting the whole process again.

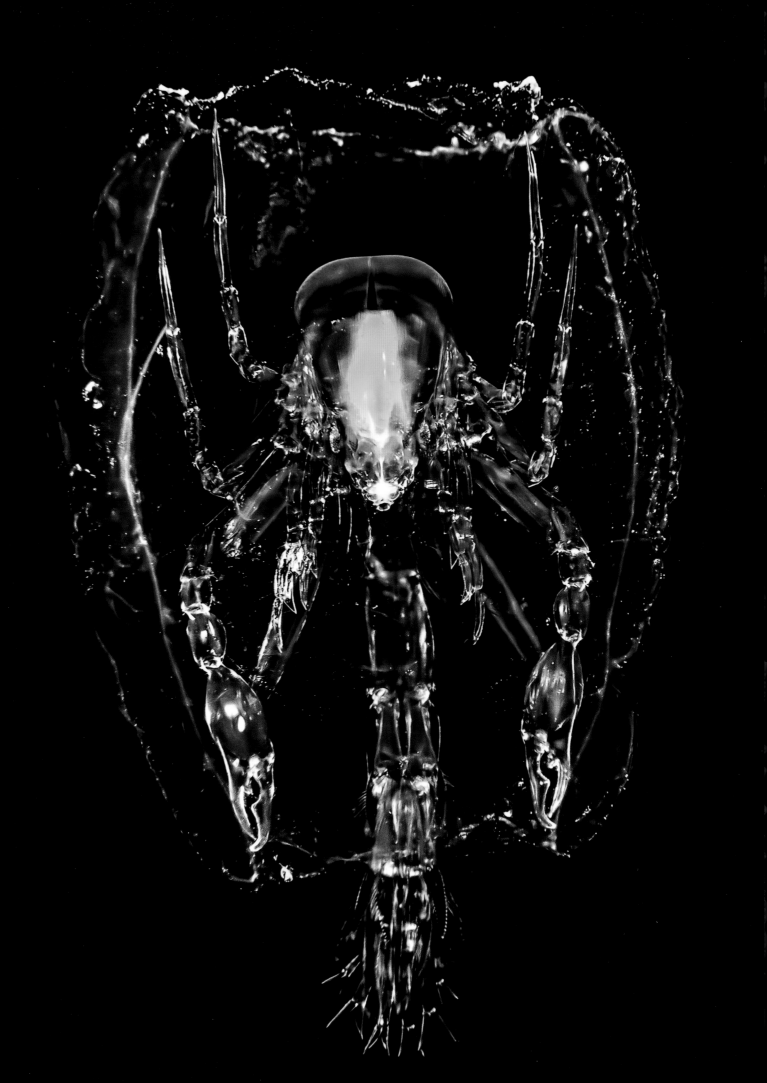

FILMING THE INVISIBLE

To film both phytoplankton and zooplankton, we teamed up with French scientists and photographers Christian and Noe Sardet, who have spent years working with these microscopic life forms at the Oceanographic Observatory at Villefranche. The big problem with filming plankton is that they are mostly invisible to the naked eye. So, to film them you need some very specialised equipment, like microscopes. But once magnified a whole new world is revealed.

Coming in a huge variety of shapes, many of these miniature organisms have an otherworldly beauty. Their alien appearance more than lives up to their names: ctenophores, siphonophores, radiolarians and *Phronima*. But getting still images of plankton is one thing; capturing footage of them as if free floating is another challenge entirely. If you put them in a tank, they just sit on the bottom or wriggle into the corners. So, the team used what's known as a Kreisel tank, which creates a flow of rotating water. The water in these tanks moves fastest at the edges so anything that's almost weightless in water – like plankton – tends to be moved into the centre, where the water is relatively motionless. It sounds simple but you are still working with a tiny depth of field where the smallest movement can throw things out of focus or shoot the subject out of frame. And there were other more modern-day problems to deal with. The samples of seawater contained large quantities of microplastic fibres (the kind that, apparently, comes from washing machines) that floated around the plankton. So all the water had to be super-filtered to remove these, and any other floating particles that would otherwise make conditions too murky to see the plankton clearly, or at all.

One of the subjects we were particularly keen on filming was *Phronima* – a group of parasitic planktonic species that preys on salps: gelatinous, jellyfish-like animals. When this parasite encounters a salp, it eats the living tissue and climbs inside, using the casing as both protection and vehicle, driving it along with its powerful legs, like some strange submarine. *Phronima* are thought to be the inspiration for the monster in the cult classic sci-fi film *Alien*. When you see these creatures up close under a powerful macro lens, you can see why. They have huge eyes and arms tipped with claws like curved daggers. 'If this thing was as large as a human child and cruised around off our beaches,' said cameraman Alastair MacEwen, 'the sea would be deserted by people.' It's just as well that *Phronima* are only 12–25 millimetres long.

It took two weeks of waiting before a *Phronima* finally appeared in one of the samples collected daily by the scientists. Even this was lucky. Alastair had once been filming on a research vessel for 45 days hoping to film one of these little aliens, and in that time only one was found. Another piece of fortune was that our *Phronima* performed perfectly in the Kreisel tank, which is not an ideal piece of equipment for strongly swimming animals like these. Despite planning and best intentions, however, one can't always guarantee a sequence makes it into the final cut of a programme. In the end, we felt it was too much of a diversion to the story we were telling about phytoplankton. These things happen – though fortunately not often.

Opposite top left **Filming plankton at the Université Pierre et Marie Curie in Villefranche-sur-Mer, near Nice.**

Opposite top right **The larval stage of a common starfish.**

Opposite middle **Sea sapphires are deep-sea copepod crustaceans that parasitise free-floating tunicates.**

Opposite bottom left **Planktonic deep-sea copepod from the Atlantic Ocean.**

Opposite bottom right **Crab larvae are constituents of deep-sea plankton.**

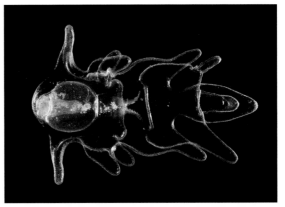

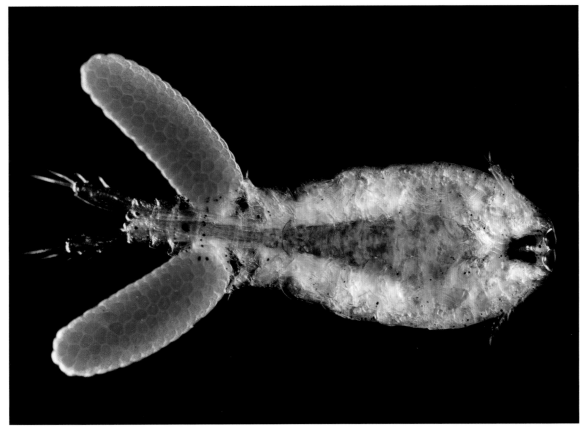

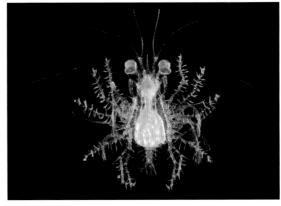

FEEDING FRENZY

Plankton are not just the most important source of oxygen in our atmosphere, they also form the basis of almost every food web in the ocean – driving the most dramatic gatherings of marine predators on the planet. For the ocean's apex predators, like dolphins, sharks, marlin and tuna, the best chance of finding food is to follow cold nutrient-rich currents, chock-full of plankton. It's here they are likely to encounter their prey – shoaling fish, like sardines, herring and anchovies, that depend on zooplankton. Once a shoal has been detected, the predators drive the fish into a ball – a 'baitball' – pushing them to the surface, which makes the fish easier to pick off. Within minutes the baitball becomes a feeding frenzy with literally hundreds of predators taking advantage: dolphins, sharks, sea lions and even some kinds of whales attack the shoal from below while, above, aerial predators like gannets and shearwaters dive bomb the fish – hitting the water at speeds of up to 80 kilometres per hour. Baitballs seldom last long – perhaps as little as ten minutes – and when the last fish has been hoovered up the predators disappear into the blue, leaving only the scales of fish raining slowly down to the deep.

Baitballs are the Holy Grail of ocean filming, with each production attempting to outdo what the previous one has captured – that's if they can even find one to film. Given that baitballs are often here one minute and gone the next, weeks can go by at sea without ever seeing one of these sudden and elusive gatherings. When open ocean shoots fail, nine times out of ten it's because the goal has been a baitball.

Below **A mega-pod of long-beaked common dolphins heads for the 'Sardine Run' offshore from Port St Johns on the coast of South Africa.**

One of the best places to try and film a gathering of feeding predators is in the cold current running along the coast of South Africa, where most years huge shoals of sardines mass (though even this event is growing increasingly unpredictable with climate change). From an underwater perspective, the best location to focus on is Port St Johns, where the water visibility is good. To increase our chances still further, we teamed up with top underwater cameraman Roger Horrocks, who lives on this part of South Africa's coastline. Having Roger close to the filming location meant that he could react at short notice. Nothing happened for months but, eventually, the perfect moment came.

During one of his trips out on the ocean, Roger spotted a big baitball, with thousands of gannets diving en masse. As they hit the water, the dive-bombing birds created a foaming churn at the surface, about ten metres wide. It was the signal for Roger to get into the water fast. He knew that even a gathering of this size might still only last 20 minutes, thirty at the most.

Once overboard, Roger was mesmerised by the sight of hundreds of birds, sharks and dolphins shooting through the corralled fish. While the dolphins picked off individuals from the edge of the baitball, sharks erupted from inside the shoal, chomping on sardines with blood streaming through their gills. As the ball of fish moved through the water, it was a real challenge for him not to get enveloped in the shoal. 'I was sweating so hard in my mask from concentration and effort that I had to stop and clear the glass just to see what was going on!' Several times he was bumped hard by sharks from behind, as they muscled in on the feeding action. Over the years, Roger has been fortunate enough to film predator feeding frenzies on sardines many

times but this one, he said, was without doubt the most intense baitball he'd ever experienced.

In addition to the underwater drama of a baitball, we also wanted to get the build-up and action above water – or *topside* as it's called in the business – and, for this, Port St Johns was definitely not the place. The good underwater visibility at this spot means it's very popular with tourists and divers, so if you're filming topside then you have the tricky issue of trying to frame out all the boats. For this *above-water* perspective, the *Oceans* team decided on Port Elizabeth, over 600 kilometres further down the coast. At this location, a combination of low underwater visibility and big seas makes it much less popular with tourists… and film crews.

Few natural history filmmakers have tried filming baitballs at Port Elizabeth, so nobody knew how difficult it was likely to be. The aim – or dream – was to film gyro-stabilised shots of dolphin super-pods from both the boat and the air, using drones. Every day, field director Raz Rasmussen and the team travelled as much as 140 kilometres offshore searching for these congregations of predators along the Agulhas Current – a narrow, but strong surface current that sweeps down from Mozambique and along the coast of South Africa.

The sea was big and the boat was small – which created daily challenges for the camera equipment strapped to the deck. Raz described it as an arms race with the sea. Every few days, the camera mount had to be braced with

Below **Common dolphins feed on South African pilchards, the main fish species in the sardine run.**

Above **Oceanic blacktip sharks let the dolphins do all the hard work, rounding up the pilchards, and then compete with them for the glut of food.**

Overleaf **Dolphins tend to corral the pilchard shoals to concentrate the fish, before ploughing into the tighter baitballs. Cape gannets plunge from above.**

another support to keep it sturdy, so that by the end it looked, according to Raz, like a game of Jenga. Sometimes the wind was so strong that the drone remained stationary when they launched it, even while flying at maximum speed. And, when returning to the boat, catching the drone in five-metre seas was a little touch and go. But the effort was worth it.

You need a certain amount of luck on every shoot, but the team was helped by a secret weapon in the shape of the boat's captain who, Raz said, was able to see things that couldn't be spotted with top-of-the-range stabilised binoculars. Thanks to the captain's eagle eyes, they saw pods of common dolphins most days – and even a group of *Pseudorca*, or false killer whales, which hadn't been seen in the area for 20 years. But the big prize was the super-pod of dolphins they encountered on one of the days at sea. A staggering 10,000 dolphins swimming in a wide line in the search for prey – the biggest group the captain, or any of the boat support, had ever seen.

SEA LIZARD

The Humboldt Current, named after the Prussian naturalist and explorer Alexander von Humboldt, is a cold, nutrient-rich current that starts its journey at the southern tip of Chile, before travelling up the coast of Peru, where it fuels one of the world's most important fisheries, and onto the Galápagos Islands. It's why you find cold-water species like penguins and sea lions in the Galápagos, even though the archipelago sits on the equator. The Humboldt's cold, nutrient-rich waters also play a crucial part in the islands' incredibly rich biodiversity. But this isn't the only deep, cold current to benefit the Galápagos.

The Cromwell Current, or Pacific Equatorial Undercurrent, flows in the opposite direction to the Humboldt. After travelling for nearly 10,000 kilometres east across the Pacific's seabed, it hits the west side of the archipelago and is driven upwards around Fernandina, the youngest of Galápagos's volcanic islands. As a result of frequent eruptions, the surface of this uninhabited island is barren and hostile. But beneath the waves it's a very different story. This cold, rich current supports huge amounts of life – both plant and animal. And that has allowed the survival of two unique land-based species found nowhere else on Earth: flightless cormorants and marine iguanas.

Flightless cormorants depend entirely on the Cromwell Current. It's provided such a reliable source of food that, in the absence of predators, the cormorants have traded the power of flight for short stubby wings, which are

Opposite **A marine iguana heads off to feed, Fernandina Island, Galápagos. Marine iguanas are the only lizards in the world to get all their food from the sea.**

Below **Marine iguanas come close to a nesting flightless cormorant, while a Sally Lightfoot crab hides under the ledge.**

more efficient when hunting underwater, since they create less drag. They can only hold their breath for a few minutes, but that's long enough to chase and catch a fish in open water, or winkle one out from the many rocky crevices below the waves.

Marine iguanas are the world's only seagoing lizard – an evolutionary adaptation that has enabled these exclusively vegetarian (or, more precisely, vegan) reptiles to exploit the Galápagos's underwater riches. Nearly half the Galápagos's population of marine iguanas live on Fernandina and tens of thousands carpet the shores, where they spend much of their day basking on the island's volcanic rock. They have highly developed salt glands that allow them to excrete the excess salt that they ingest while foraging at sea and, every so often, they spray jets of salty snot over their sunbathing neighbours. Nobody seems to mind.

When sufficiently warmed up, the iguanas head to the sea, leaping into the surging waves. The larger iguanas must swim beyond the breakers to reach the best grazing – a feat in itself given the force of water smashing against the rocky coast. But the waves are by no means the only issue the iguanas must deal with. Like all reptiles, marine iguanas struggle in the cold,

Above **Marine iguanas can stay submerged for up to 30 minutes, but more usually feed in much shorter bouts, during which they scrape algae from the rocks.**

OCEANS

162

and the temperature of these waters can get as low as 11°C (very different from their ideal body temperature of 35–39°C). If an iguana stays too long in the chilly sea, its muscles will seize and it'll drown. So the clock is ticking the minute an iguana launches into the water. It has just 30 minutes to reach the grazing grounds, feed and get back: an operation it must complete every day of its life. But, naturally, it's well designed for the job – it can hold its breath for the full duration of the dive; it has a flattened tail, which makes swimming more efficient; and its flat face and specially designed teeth make it easy to crop the algal growth it depends on for survival.

When it has had its fill of seaweed, each iguana makes a beeline back to the safety of the hot volcanic rocks – often running the gauntlet of playful sea lions, who like to amuse themselves by grabbing the tails of the swimming lizards. These hold-ups could make the difference between life and death, since an iguana is losing vital heat with every minute in the cold water. If its body temperature drops too far, it won't have the energy to push through the surf and scale the rocks to get to the basking zone above. Once out of the water, however, it will warm up quickly, helped by its black skin – the perfect colour to absorb heat from the sun.

BIG WAVE SURFER

Marine iguanas are one of the Galápagos's most iconic species. They are found nowhere else on Earth and their behaviour is unique in the lizard world. It's hardly surprising, therefore, that they have featured in numerous wildlife documentaries over the years. Despite this, we were still keen to include them in the *Oceans* episode. They were, after all, a perfect illustration of how currents can bring life to an otherwise barren part of the world. But the question was how to bring freshness to the sequence. Enter cameraman Richard Wollocombe. Richard has spent the past 25 years working in the Galápagos, and no wildlife filmmaker knows these islands better. In fact, if you've seen a sequence shot in the Galápagos, the chances are it was wholly or partly shot by Richard (the famous snakes and iguanas sequence from *Planet Earth II* is a case in point). And Richard did indeed have a new idea for the iguanas – a part of their behaviour that had never been properly featured in a wildlife sequence: their journey through Fernandina's powerful surf on their way to the feeding grounds. It didn't take him long to convince Ed Charles, the producer, to give it a go.

One of the most frequently asked questions for people working in wildlife films is, 'Hasn't it all been filmed before?' Understandable, perhaps, given how many natural history films have been made over the years. But even leaving aside the power of new technology to show familiar things in a completely fresh way, it's amazing how many un-filmed species and behaviours are

Below left An inquisitive sea lion plays with a marine iguana as it tries to return to land. Once in the cold water, the iguana has around 30 minutes of activity before it has to return to base. Any longer and its muscles will seize and it will drown. So playful sea lions are not exactly welcome.

Below Galápagos wildlife is remarkably tolerant of humans, which makes filming that little bit easier. Here, cameraman Richard Wollocombe is filming marine iguanas on Fernandina Island.

Opposite Richard Wollocombe on the marine iguana shoot, Fernandina Island, Galápagos.

still out there – and the surf-swimming iguanas are a case in point. It's why each new major BBC landmark natural history series seems to capture people's imaginations.

There was, of course, a good reason why nobody had filmed the iguana from inside the surf before – it was potentially very hazardous. Richard had seen iguanas swimming through the waves many times before but always from a safe distance. Now it was time to get up close and personal. As he said, 'It took a while to understand how the waves interacted with the topography of the ground, and where the optimum areas were to film the drama without exposing ourselves to unacceptable risk.'

Once all health and safety issues had been considered and assessed, Richard and his two safety divers, Rafael Gallardo and Juan Carlos Banda, got into position. The first problem that arose – and something that couldn't be predicted from shore – was that the iguanas were clearly nervous of swimming anywhere near the divers in the surf. Animals in the Galápagos are famously naïve about people – never having cause to feel afraid of us – which means getting close to the islands' wildlife is, on the whole, relatively straightforward. Not this time. Perhaps they just didn't recognise Richard and his buddies as humans, given that they were in full scuba gear and wearing crash helmets to protect against contact with the rocky coast. Whatever the reason, it was surely an indication that what we were trying to film was new.

The second issue was one that the team did feel well prepared for – the difficulty of being in full diving gear, with tanks of compressed air, and then pushing a cumbersome underwater housing while trying to maintain their

position in the surging waves. But imagining this and actually doing it are two very different things. The team knew that being caught full-on by a wave could result in serious injury, so Richard and his dive buddies had to be ready to take evasive action irrespective of what the iguanas were doing. Missed opportunities were inevitable, as Richard acknowledged: 'Often the best moments with the iguanas were happening precisely when the waves were at their most dangerous, so if I wanted to avoid being ripped to shreds on the rocks I had to stop filming. But getting any kind of shot was tricky when, on the one hand, you're trying not to scare your subject, while, at the same time, not terrifying yourself by the thought of being caught in a powerful wave.'

For a while it looked like the idea might not work. But, to use an old wildlife filming cliché (and one that you'll hear at the end of almost every

Below Marine iguanas on Fernandina must plunge through the surf to reach their feeding sites offshore.

Making Of or *Behind the Scenes* section), 'perseverance eventually paid off' – perseverance and the usual combination of luck and judgement. 'When one came towards us,' said Richard, 'we would dive down to the bottom, and with one hand I would hold the camera and film and with the other grab on to a rock and hope that the wave didn't rip me up and throw me into the turbulent chaos of the white water.' The experience gave the crew a new appreciation for the marine iguanas, who have to battle the waves and current every day of their lives just to get a meal. The hazards of being thrown against the jagged volcanic rock by large waves are as real for the marine iguanas as they are for a film crew – and iguanas do regularly die this way.

BIG CURRENT, SMALL CUTTLEFISH

One of the largest and most powerful currents on Earth occurs in the Indonesian Archipelago and is known, rather boringly, as the Indonesian Through-Flow (ITF). It's the result of a difference in sea level between the Pacific and the Indian Ocean. Every second, the ITF carries 15 million cubic metres of warm water from the Pacific to the Indian Ocean, where it eventually makes its way to the Atlantic Ocean. This current performs a vital role in the global climate system because the heat it carries from the tropical Pacific to the Indian Ocean provides the energy for the Asian and Australian monsoons – but it's also responsible for creating the richest coral ecosystem on the planet: the Coral Triangle, covering an area of 5.7 million square kilometres.

Most of the ITF flows through the Macassar Strait between Borneo and Sulawesi before spreading around and through the Indonesian Archipelago. Bringing together species from two different oceans, this region is home to a third of the world's reef fish and more than 500 reef-building corals, making it the most diverse marine area on Earth.

Evolution has run riot in these productive waters – giving rise to some truly bizarre animals, from pipefish and boxfish to frogfish. But surely the oddest is the tiny flamboyant cuttlefish – a cephalopod group that also includes squid and octopus. Like most other cuttlefish, it's a quick change artist, able to blend into its surroundings, which is particularly useful when stalking its prey. Unlike other cuttlefish species, though, its extra-small cuttlebone means it can't

swim for long. Instead, it prefers to crawl along the ground. The cuttlefish's *flamboyant* moniker comes from the flashes of bright yellow and pink it makes when startled, or trying to attract a mate.

The five-centimetre males are significantly smaller than the females (a trait known scientifically as sexual dimorphism). A male's role in the breeding process, once he's won the object of his affections over with sufficient flashes and dazzles, is only to deposit a packet of sperm in her mouth – though given the size difference between the sexes, this can look like a daunting prospect. The female will then lay their fertilised eggs in a safe place, away from predators, like the underside of an abandoned shell – gluing anywhere between 15 and 50 eggs to the structure. The huge current that brings so much life to the Coral Triangle now washes the eggs with clean, oxygenated water. After two weeks, the babies are ready to hatch, emerging as exact replicas of their parents, though very tiny. Smaller than a human fingernail, the hatchlings now make the most of the flow of water, allowing themselves to be carried with the current to colonise new parts of the reef.

Above **After two weeks of incubation, a baby flamboyant cuttlefish is ready to hatch from its egg. At this age, the cuttlefish are smaller than the size of a human fingernail.**

Opposite **Flamboyant cuttlefish have a dark brown base colour, but can switch in an instant, adopting an endless repertoire of camouflage colours when stalking prey.**

STAYING ON SCRIPT

The location we chose to film the flamboyant cuttlefish was Lembeh, off the coast of Sulawesi, which is world famous for *muck diving*, a type of diving that is done in fairly flat muddy or sandy areas. In addition to the disco flashing, flamboyant cuttlefish, Lembeh is home to an amazing array of crazy-looking animals, including harlequin shrimps, slipper lobsters, blue-ringed octopuses and orangutan crabs. But all our effort went into the cuttlefish, which no one has ever filmed a whole sequence on before.

Despite the cuttlefish's tiny size, finding subjects to film was, fortunately, not an issue. Prior to the shoot, local guides spent ten days scouting the area and by the time the team arrived the territories of several cuttlefish had been identified.

Filming underwater is a complex business at the best of times and it's even more difficult when your subject is just a few centimetres long and lives in a place with a strong current. To maintain their position, cameraman Hugh Miller, camera assistant Dan Beacham and field director Raz Rasmussen had to anchor themselves to the ground with large sticks pushed into the mud. Getting the range of behaviour the team were after required a lot of waiting around, so both Hugh and Dan dived using rebreather systems, which enabled them to spend three hours at a stretch underwater – as opposed to the hour or so you'd get with scuba. Rebreathers also have another significant advantage over scuba gear. They create much less disturbance since they don't release

Above **Cameraman Hugh Miller filming flamboyant cuttlefish in Lembeh, Indonesia. The two rock-like lumps in the sand are female flamboyant cuttlefish.**

any air bubbles, which often scare or distract marine creatures. (The reason rebreather systems don't release bubbles is because the compressed air is completely recycled within the unit.)

Over the course of a day, the team would spend a total of six hours underwater – not dissimilar to a topside cameraperson sitting in a hide. Naturally, you can get hungry and thirsty during long stints of filming, so Hugh and Dan would bring gel pouches (like those used by athletes), which they would suck on underwater. If things got very slow Hugh would bring out his waterproof Kindle and settle down on the seabed for a good read.

The old joke when filming wildlife is that animals never read the script, so it's not unusual to have to adapt what you planned in the office in order to fit in with whatever behaviour you do actually see – and that's if the animals decide to make an appearance in the first place. Our flamboyant cuttlefish, however, really broke the mould and did literally everything that was required of them. It was as if they really had read the script. For Hugh and Raz, they were the most compliant animals that either of them had ever filmed. The team were also able to disprove the accepted science that females die after laying their eggs, since ours didn't. (Most of the work done on flamboyant cuttlefish has been in captivity where conditions are naturally different.)

The only disappointment of the shoot was the amount of plastic and other litter the team encountered across the dive site. As the current ebbed and flowed, Raz said, 'countless pieces of plastic zipped past our eye-line whenever we looked up into the water column.'

Below **Cameraman Hugh Miller and assistant cameraman Daniel Beecham filming flamboyant cuttlefish, Lembeh.**

The ground wasn't much better. When they dug out the sand to settle the camera into position, all manner of man-made stuff appeared, from Lego men to crisp packets and bottle tops. One day Raz took a canoe along the coast and he said some of the beaches were almost knee-high in plastic rubbish.

THE PULL OF THE MOON

Three hundred and eighty-three thousand kilometres from Earth, through pure cosmic good fortune, we are orbited by a moon. It's thought to have been created from the debris resulting from a vast asteroid colliding with Earth, early in our planet's history. The moon is only one-hundredth the mass of Earth but that, and its proximity to this planet, means it still has a strong and consistent gravitational pull on our world. And the significance of this is revealed approximately every 12 hours along our coastlines in the rhythm of the tides. (It's worth stating here that the moon is actually slowly moving away from Earth. When the moon formed some four to five billion years ago, it was over 19 times closer than it is now, resulting in a much greater gravitational pull and correspondingly higher tides.)

With every full rotation of the Earth, there are generally two high tides and two low tides. The gravitational pull of the moon causes the ocean to bulge at two points, both of which are aligned with the moon – or, to be more precise, at the nearest point to the moon and on the exact opposite side. The reason for the latter is because the moon's gravity also stretches and squashes the Earth (as a soft rubber ball would be when squeezed with thumb and fingers). Low tides occur halfway between those two points. And as the Earth rotates, the high and low tides move around the planet accordingly.

OCEANS

However, although the gravitational pull is the same across the planet, its impact on the coastline is not. Continental landmasses disrupt the movement of water so that some places get higher or lower tides than others. The average tidal range is about a metre, but the Bay of Fundy, in Canada, has a tidal range of around 13 metres – the greatest on the planet. Winds can also exaggerate tides. Strong offshore winds, for example, can push water away from the coastline, accentuating a low tide. Onshore winds can do the opposite, making a high tide even higher. Storms and hurricanes – resulting from low pressure systems – can dramatically increase the effects of a high tide, as the city of New Orleans tragically discovered in August 2005 when Hurricane Katrina led to a storm surge of over eight metres.

Tidal ranges also vary across the month. The highest are the spring tides (confusingly the name has nothing to do with the seasons, but because they *spring forth*), which occur twice a month on the full and new moon, when both the moon and sun align with the Earth. This alignment causes extra gravitational pull – even though the sun is nearly 400 times further away from Earth than the moon. The opposite of a spring tide is a neap tide, which occurs a week later when the sun and moon are at right angles to one another, in respect of the Earth. During neap tides, which also happen twice a month, the gravitational pull of sun and moon cancel each other out, causing a smaller tidal range.

DEEP-DIVING DUCKS

Tidal currents don't have much impact on the open ocean, but they can create currents of up to 25 kilometres per hour when they flow in and out of narrow bays and estuaries. Nowhere experiences the power of the tides more violently than Norway's Saltstraumen Strait. Here, every six hours, nearly half a billion tonnes of water are forced through a channel just 150 metres wide. As it moves through the narrow gap, the water picks up speed – resulting in the strongest tidal current in the world. Most animals would be swept away. Not eider ducks. The eider is one of the few ducks to totally depend on the ocean for survival, and it takes even these powerful currents in its stride. The colourful eiders – one of the world's largest species of duck – are drawn here by the great feeding opportunities underwater: in particular, the extensive beds of mussels anchored to the seafloor. The currents are perfect for the mussels, too, who filter out the plankton from the fast-moving currents.

Sticking together in a large group, the eiders appear to bob around in the current like corks. But when it's time to go hunting for mussels they simply do what ducks do best – duck dive. Eiders are especially talented at this, being able to dive to depths of over 40 metres. Once below the surface, they push themselves to the bottom with among the strongest legs in the duck world. When they reach the beds of mussels, they winkle them off the rocks with their specialised wedge-shaped bills. Then the mussel is swallowed whole, shell and all. In the Saltstraumen Strait, the eiders have these molluscs to themselves and each duck will eat hundreds a day – sometimes more. One eider, in fact, was recorded eating 1,600 mussels in a single day – as well as 15 crabs and six starfish.

Above **A gathering of eider ducks – males are white, females brown – in Norway's Saltstraumen Strait. They are one of the few animals able to handle Saltstraumen's powerful currents, where they forage for mussels on the seabed.**

Opposite top **A whirlpool churns the sea in Norway's Saltstraumen, a small strait that experiences the strongest tidal currents in the world.**

Opposite bottom **Three eider ducks harvest mussels from the seabed. Each eider can eat hundreds of mussels a day, which they swallow whole.**

QUACK, QUACK, OOPS!

Above **The crew on the road in the Norwegian Arctic during the eider duck shoot.**

Opposite **Filming aerials of the Saltstraumen Strait with a drone.**

Filming ducks sounds easy, particularly when compared to other wild animals. After all, the local town near the location had a population of eiders that have been attracting visitors for over 30 years. Raz Rasmussen was even told that many of the eiders could be reliably hand fed. When the team arrived, however, the place was pretty much duckless. As mentioned in a previous chapter, natural history filmmakers are often told by the locals that the animal and behaviour they've come to film would have been much easier to find and better if done earlier (as in a previous week or month) but here, at least, it was true… We really should have been filming in Saltstraumen the year before. Numbers of eider ducks have, apparently, been plummeting across Norway – why, nobody seemed to know. We did eventually find a population to film underwater, although, for some of the team, this involved a 13-hour car journey over a snowy mountain, driving behind a safety vehicle with tank tracks and snowplough, going at seven kilometres an hour.

HIDE AND SEEK

If the tidal currents of Saltstraumen Strait move like a fast-flowing river, those of the Bahamas are more like a slowly filling bath. Here, the wide, shallow sandbanks mean the tides move gently over the seafloor. Nevertheless, while the speed may not be dramatic, the toing and froing of the ocean turns what would be a sandy desert into a rich underwater habitat. The sand is home to razorfish and garden eels who pick nutrients out of the gentle, twice-daily current. It would seem a perfect life were it not for the pods of predatory bottlenose dolphins. When these predators come into view, both eels and razorfish are quick to disappear under the soft, golden sand. A good hiding place, you'd think, but bottlenose dolphins aren't easily fooled by the vanishing trick, and they have a trick of their own – echolocation. The dolphins make a sweep of the sand, scanning it with a series of high-pitched clicks. The returning sound tells them where a razorfish is hidden. When one is detected, the dolphin digs it out of the sand with its snout. For a hard-to-reach razorfish, the dolphin blows jets of water into the sand to expose it.

Right A dolphin blows jets of water into the sand to expose a hard-to-reach razorfish.

Above left **Razorfish** emerge from the sand to feed on food particles brought in by tidal currents. At the first sign of danger they hide under the sand.

Above middle **Two** bottlenose dolphins on the hunt for razorfish. They use echo-locating clicks to find the hidden razorfish.

Above right **Got one!**

BETWEEN THE TIDES

The planet's coasts support thousands of species and many of them are adapted to the ebb and flow of the tides. Barnacles are tidal specialists and a familiar sight in that zone between land and sea. Glued tight to rocks, these crustaceans can easily cope with the battering of waves as the tide floods in. When the sea goes out, and the barnacles are left high and dry, they get through these tough times by holding water in their shells and battening down the hatches.

For those that can't afford to dry out, there are the intertidal pools – rock pools – where all manner of creatures can fairly safely hang out until the incoming tide replenishes the pools with nutrients and fresh, oxygenated water. For many children, rock pools are their first glimpse of the ocean's incredible diversity and a chance to get up close to some of its inhabitants. However, few people would get too excited about one of the most common rock pool species – blue-green algae. Yet these single-celled organisms, which form into filaments and colonies, have been around for over a billion years, making them one of the earliest life forms on the planet.

Some animals time their breeding around the tide timetable. It's not just sea turtles that nest on the highest tides; fish like grunion and capelin do too, and large shoals of these small silver fish appear in the shallows when it's time to spawn. A few days after the highest tide, both males and females beach themselves in a synchronised spawning event, rolling in on waves to reach the high-water mark. The fertilised eggs are buried in the dry sand, away from aquatic predators, and there they will develop until the next very high tide washes the eggs back into the sea, whereupon the developing larvae immediately hatch. If the tide doesn't quite reach the right spot, the eggs can even delay hatching for another two or four weeks until the next big high tide. How the grunion and capelin know exactly which tide to spawn on nobody can say for sure, but scientists suspect that they have a *circatidal* clock (similar to a circadian clock that follows day length), which enables them to detect minute changes in the gravitational pull of the moon.

The mouths of large rivers often form another familiar and important tidal habitat. Estuaries are where sea and freshwater meet and they're vital feeding grounds for huge numbers of birds, like oystercatchers, knot, dunlin, avocets and sandpipers. These wading birds move with the tides, using specialised bills to probe the exposed mud for prey such as mussels, cockles, clams, ragworms and limpets.

Opposite **Giant green sea anemones, goose barnacles and ochre sea stars are exposed at low tide in Olympic National Park, Washington.**

Overleaf **At Snettisham, on the east coast of England, oystercatchers, knots and other waders arrive on the mudflats at low tide to feed.**

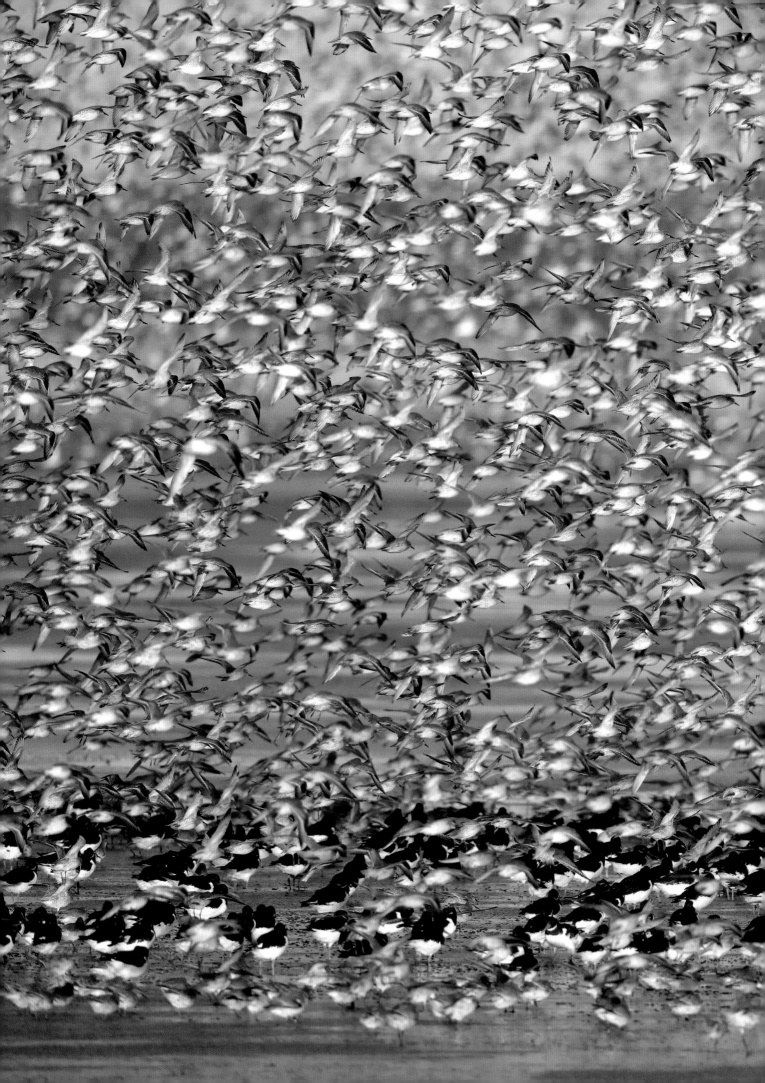

SALTWATER FOREST

Perhaps the most important tidal ecosystems in the world are mangroves – essentially estuaries with trees growing out of the mud. These largely tropical, tidal forests are some of the most productive habitats on Earth, as well as the toughest. The daily flooding would kill most trees, but mangroves thrive in these conditions. They are the only trees capable of surviving in saltwater – up to 100 times saltier than most other plants can tolerate. They do this either by excreting the excess salt from their leaves or by filtering it out as the seawater enters their roots. Some mangrove trees even store the salt in older leaves or bark, so that when the leaves drop or the bark is shed the salt goes with them. The freshwater that remains after the salt has been extracted is stored in tough, waxy leaves which minimise evaporation in the hot tropical sun.

Below **Two juvenile lemon sharks swim through a channel on the edge of a mangrove swamp, Bimini, Bahamas.**

Breathing is another challenge mangrove trees have had to overcome. In normal forests, trees can take in oxygen from gases trapped in the surrounding soil, but that's not possible in mangrove forests, where the muddy soils are very low in oxygen. So, to deal with this problem, the trees grow aerial roots, called pneumatophores, which take in oxygen from the atmosphere. It's this network of aerial root systems that makes mangrove forests so distinctive.

As well as providing a daily flow of nutrients, tidal currents also help to spread the trees' seeds. Highly buoyant, the arrow-like mangrove seeds have been known to drift in the ocean for more than a year before taking root. When the seed reaches brackish water – a mixture of fresh and salty water and the ideal growing conditions for mangrove trees – it becomes vertical so that as the tide goes out it sinks into the mud with its roots pointing downwards.

Mangroves are the foundation of a coastal food web supporting thousands of species from invertebrates to birds and mammals, including oddities like pygmy three-toed sloths, proboscis monkeys and mudskippers. Like something out of the topsy-turvy world of *Alice in Wonderland*, these bulbous-eyed fish can climb trees, which they do using their pectoral fins. And by storing water in their mouth and gill chambers, mudskippers are as happy out of the water as in it. Mangrove forests are also crucial nursery grounds for juvenile fish, which take shelter amongst the network of exposed mangrove roots. When the tide floods in so too do other larger fish.

Below **An Atlantic mudskipper uses its pectoral fins to crawl across a tidal mudflat amongst mangroves in the Bijagós Islands off the coast of Guinea-Bissau, West Africa.**

Above **A newly born lemon shark pup swims away from its mother in the Bahamas, West Indies.**

In the Bahamas, predatory stingrays and juvenile lemon sharks ride the incoming tide to hunt in the mangroves. When the water is at its highest, even adult lemon sharks can reach the forest. But these large, three-metre predators don't come to the mangroves to hunt. They are females that have come to give birth, returning to the very place they were born. Lemon sharks can have as many as 18 pups and as soon as they're born the female heads back out to deeper water, to avoid being trapped by the falling tide. The newly born sharks, however, make for the security of the mangroves. They're small enough to swim deep into the labyrinth of trees – though if they're not to be left high and dry themselves they must find one of the forest's permanently flooded pools: a shark nursery that will be out of reach of predators. The young sharks will spend the first two years of their lives in these pools learning what it takes to be one of the ocean's top predators.

HIGH AND DRY

It's not just female lemon sharks that need to keep an eye on the falling tide; film crews do too. To get to their filming location in the mangroves near Bimini, the team needed to navigate the same route through the labyrinth that the juvenile sharks used – waterways that were only accessible by boat for an hour or two either side of high tide. If they headed out too late the window of opportunity would be gone, meaning a day's filming would be wasted. If they started on the return trip too late, however, they would be left stranded in the mangroves – something that Ed Charles, the producer, was naturally keen to avoid. As Ed said, 'If that happened, there would be five adult men stuck on a tiny boat, with no shelter, having to wait for ten hours, mostly in the dark, for the tide to come back in.' Clearly no one wanted to risk that happening, but you can lose track of time when you're focused on filming, and there was one moment on the shoot when the ten-hour wait nearly became a reality, as Ed recalled later. 'After quickly jumping off the boat, it took all of our strength to half lift/half push the boat off the sand bar we'd got stuck on to get it into deep enough water and safety. But it was a very close-run thing.'

Below **Cameraman Duncan Brake and producer Ed Charles film juvenile lemon sharks in the mangroves, Bimini, Bahamas.**

And there were other, more dangerous matters to be aware of while filming. To get the cameraman in place often meant jumping out of the boat and then slowly pushing it through the water to the required location. No problem about that. But the team knew that lurking in the sand and seagrass were highly camouflaged stingrays, armed with barbed, venomous stingers. As a precaution, the team adopted a technique they called 'the shuffle' – which involved sliding their feet gently along the bottom. There was always the chance of bumping into a stingray but this method of walking meant they wouldn't actually stand on one, which was the real danger. People struck by these relatives of the shark have described it as the worst pain they've ever experienced. So, the *shuffle* was the best way to avoid what might otherwise have ended up as the *hop and scream.*

While nobody wanted to get close to a stingray, the opposite was true for the young sharks, who were the stars of this shoot. Fortunately, this didn't turn out to be a difficult challenge. To a shark, splashing could mean an injured fish and so as soon as the team started to move around the pool they were investigated by all the young sharks in the vicinity. Once curiosity had overcome fear, there would be as many as 20 young lemon sharks around the crew, swimming through their legs and sometimes bumping them with surprising force. Slightly unnerving, perhaps, but these fish were harmless. There was only one shark they worried about – an individual the crew nicknamed Evel Knievel, who, according to Ed, was rather bitey. It was Ed's job to keep young Evel at bay, while the cameraman was focused on filming – making sure, for instance, it didn't take too close an interest in unprotected fingers. Thankfully, it was easy to see him coming as he had a distinctive black smudge on his head, so Ed could intercept him with a protective stick and then gently push him away.

Opposite **Cameraman Doug Anderson films stingrays, Bimini.**

Below **Filming a juvenile lemon shark amongst the mangrove roots in Bimini.**

MANTA BONANZA

All life at the coasts moves to the daily rhythm of the tides but for some animals it's all about waiting for just the right tide: the spring tides, the highest tides of the month. On one particular reef in Micronesia, it's the cue for a very special event. Thousands of resident surgeonfish begin to gather, waiting for dusk. Then, in a synchronised movement, both males and females swim up into the water column, while at the same time releasing billions of eggs and sperm. Breeding at this particular time, when the tide is at its highest, means the fertilised eggs will be carried out by the moving water to the open ocean and away from reef predators. Or most of them. What makes this story even more spectacular is the sudden presence of manta rays.

The mantas spend the year moving between remote coral islands but it's only when the tide is at its highest and the surgeonfish gather that they appear on this reef. Nobody knows how the mantas are so perfectly in tune with the cycle of the tides but they're here whenever the fish spawn – often arriving within half an hour of the start of the action. When the surgeonfish release their eggs, the mantas cruise through the milky water like shadowy apparitions, filtering the eggs with specialised gill plates. If they'd arrived even an hour after spawning there would have been nothing left. Nevertheless, the surgeonfish release so many eggs that most will be carried out into the open ocean before the mantas have a chance to eat them all.

Filming this tidal dance of surgeonfish and mantas was only possible through close collaboration with the scientist that discovered the site. But there was an important condition attached to filming here – we were sworn not to reveal where the event takes place. There's good reason for the secrecy: if the location was made public, this unique site would soon get swamped with recreational divers and snorkelers. Fifty manta rays swimming in water

Right **A manta ray feeds on the recently released eggs of surgeonfish, on a reef in Micronesia. The manta's timing is perfect – arriving just as the surgeonfish are spawning, which the reef fish do on the highest tide of the month.**

Below left **A manta ray glides over the reef in Micronesia.**

Below **A manta ray circles around a shoal of surgeonfish, waiting for them to spawn.**

sometimes less than a couple of metres deep would be a major attraction and the attention could easily scare away the rays. What I can tell you is that Raz Rasmussen said the scientist and her husband – Julie and Jason – couldn't have been more supportive and generous with their time. (Also the boat Julie and Jason used to carry out their work was the nicest Raz had ever filmed from!) At one point, Raz lost half of his front tooth and Jason, who happened to be a dentist, spent a lot of time giving him an even better tooth than he'd had before. It was clear to Raz that the future of this site, and the mantas that use it, are in good hands.

Below left **Spawning** surgeonfish. Males and females shoot to the surface and, in a synchronised movement, release billions of eggs and sperm into the water. The fertilised eggs will be carried on the tide and into the open ocean.

Below right A group of manta rays feed on surgeonfish eggs. They filter them out using specially adapted gills.

Overleaf Three manta rays filter eggs from the water following a surgeonfish spawning event. These mantas spend the year moving between remote coral reefs in Micronesia.

BREAKING WAVES

Nazaré, Mavericks, Pipeline and Teahupo'o – these names may not mean anything to you. Unless, that is, you're a big-wave surfer – in which case, they're the stuff that dreams are made of. Surfing waves at Nazaré, on the Portuguese coast, can rise to an extraordinary 30 metres – the biggest on the planet, though only a handful of surfers have the skill and courage to tackle them at their most fearsome. The geography of the coastline at all these locations plays a part in creating the monster waves, but each and every wave, wherever it breaks, is the result of something that happens far out to sea.

Winds blowing over the sea stir the surface into swells, or surface waves, which, with enough energy, can travel across whole oceans. In effect, energy is being transferred from the wind to the wave. When the swell reaches land, the decreasing depth of the water results in a breaking wave – typically this happens when the water depth gets down to one and a half times the wave height. The size of the breaking wave depends largely on the amount of wind that started the swell in the first place. The more powerful the storm, the larger the swell, which is why big-wave surfers keep a close eye on weather systems out in the open ocean. Essentially, it's an advanced warning system for great surfing. It was something I once discovered for myself while trying to film someone surfing a large barrel wave breaking on a small Micronesian island called Pohnpei. For two weeks, we'd waited for the right conditions during the, supposedly, perfect time of the year when the biggest waves were most likely – but the sea remained frustratingly calm day after day. There were waves but nothing remotely challenging to the big-wave surfer we were working with – and certainly not even a glimpse of a barrelling wave. Eventually, we abandoned the shoot and went home.

Above left **A shoal of hardyheads in the shallows of Lizard Island, Australia.**

Above **The fish leap out of the water in an attempt to avoid predators.**

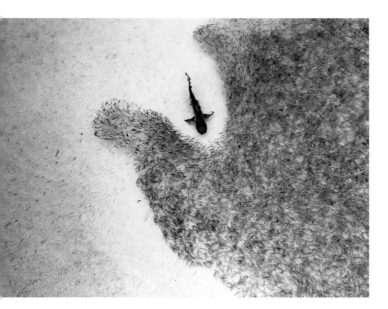 

Above **A blacktip reef shark cruises through a large shoal of hardyheads in Lizard Island.**

Above right **Two blacktip reef sharks beach themselves in an attempt to catch hardyheads.**

A few months later, I got a call from the manager of the surf camp we'd stayed at in Pohnpei saying a large swell had been detected somewhere out in the Pacific Ocean and it was due to hit the island in three or four days. I immediately assembled the team, and we all headed out to Pohnpei as fast as could be arranged. The morning after we arrived, large beautiful barrel waves came rolling over the reef and we got the perfect shots in just a couple of hours. In all my years making wildlife documentaries, it remains my shortest and most successful shoot.

But it's not just surfers that benefit from waves. Along with tides and currents, they play a crucial role in bringing nutrients to marine animals along the coast. Even the smallest waves stir the sand, exposing food that may otherwise remain hidden. In the shallows of Australia's Lizard Island, for instance, the gentle wave action can lead to a feeding frenzy in less than half a metre of water.

Here, shoals of hardyheads (small baitfish) pick off nutrients where the breaking waves kick up the sand. Their presence attracts predatory trevally, which dart into the shoal, forcing the hardyheads to take cover in the waves themselves – out of reach of the trevally. It's a good tactic until blacktip reef sharks arrive and team up with the trevally. Now, when the trevally attack and the hardyheads hide in the waves, the sharks launch themselves into the surf line and onto the beach, grabbing any fish left flapping on the sand. The hardyheads that escape swim back to sea and into the mouths of the trevally. All the activity attracts herons and gulls, which are quick to take advantage of the confused hardyheads. It's an ocean baitball in less than 50 centimetres of water.

PENGUIN COMMUTE

Opposite On the wave- and wind-swept Falkland Islands, rockhopper penguins climb the steep and rugged cliffs from the sea to their nest sites at the top.

Below A rockhopper penguin with a recently hatched chick. Both parents share the chick-rearing duties, though incubation is largely down to the males.

The more tempestuous the seas, the richer they are. The Falkland Islands lie in some of the stormiest waters on Earth and are great hunting grounds for rockhopper penguins, who also raise their young on the rocky coastline. Once the eggs have been laid, it's the males who do the incubating, leaving the females to go off hunting for krill, squid and fish – returning just as the chicks hatch. But to reach the colony, the females must scale steep rocky cliffs. Now the waves that made feeding so good become a major obstacle. The females have no choice but to battle the surf zone. Timing is vital. If they go too early they could be dashed against the rocks; too late and they will be dragged back out to sea. Hooked claws help them get a purchase on slick rocks, but it's not always enough. Time and again the little penguins are pulled back down the steep slope by the huge breaking waves. But they can't afford to give up. With hungry chicks waiting – and an equally hungry partner, who may not have eaten for several weeks – the new mothers must keep trying until they succeed. And, amazingly, they do.

The rockhoppers' journey home must surely be one of the most dangerous commutes on the planet. It is certainly no mean feat to film, as producer Ed Charles discovered. 'To get the shots we wanted meant donning full climbing gear, roping up to a parked Land Rover left in gear and on chocks, and abseiling down a very slippery, guano-covered rock face. This got us into position about 20 metres above where the penguins were beginning their ascent, and when the storms rolled in the waves slammed into the cliff, sending spray nearly 30 metres into the air and down on to us. Seeing the incredible power of the ocean here, even from the relative safety of our ropes, it was astonishing that any penguin survived the ordeal.' Rockhoppers may be one of the smallest penguins, but they're definitely the most plucky.

VOLCANOES

HELL ON EARTH

Most people thought it was the sound of distant cannon – even those who, unknowingly, were more than 2,400 kilometres from the source. Those who heard it were so convinced that the noise was that of an invading army, pirates or other aggressors that, in towns and communities all over the region, men were dispatched to find out what was going on – or, in some cases, to deal head on with whatever was causing the conflict. The date was 10 April 1815.

As it turned out, what everyone had actually heard were the explosions and eruptions of Mount Tambora, a volcano on the island of Sumbawa in Indonesia. To people living in the vicinity at the time, the scale of Mount

Above **Plosky** is a small shield volcano in the volcanic complex of Tolbachik, in the northern part of the Kamchatka Peninsula, Russia.

Tambora's final eruption was beyond imagining. Today, it is a true legend in the history of eruptions and the largest ever recorded by humans. In the Volcanic Explosivity Index (VEI) – the equivalent of the Richter scale for earthquakes – Tambora ranks a 7, out of 8. (Krakatoa in 1883 was a 6. The most recent 8, the eruption of New Zealand's Taupo volcano, is thought to have been over 25,000 years ago.)

Tambora's final eruption blew over 1,000 metres off the height of the volcano, destroying all the island's vegetation. Pyroclastic flows – immense avalanches of ash, hot rocks and equally hot gas – tore down the mountain at over 160 kilometres per hour and into the sea, setting off a tsunami of four-metre-high waves, which inundated islands and swamped coastal regions. Huge floating rafts of pumice, some up to five kilometres across, trapped ships in harbour.

The atmosphere was so thick with volcanic dust that it blocked out the sun for days. According to eyewitnesses, it was so dark that, at times, you could barely see your hand in front of your face. Ash rained down for weeks – in some areas reaching a depth of over a metre. Roofs collapsed under the weight of the debris – even houses hundreds of miles from the eruption were destroyed.

The death toll of people living on the islands of Sumbawa, Lombok and Bali is estimated at over 70,000. Approximately 12,000 people died instantly from pyroclastic flows, the rest through starvation and disease, making it not just the largest eruption in recent human history, but also the deadliest. And the story doesn't end there.

The eruptions were so powerful that they ejected huge plumes of fine ash, dust and volcanic gases – mostly sulphate aerosols – into the stratosphere, some 20 kilometres above the surface of the Earth. As the cloud circled the planet, it reflected the sun's heat back into space and altered the climate on the other side of the world.

The following year, 1816, became known as 'the year without summer' in the Northern Hemisphere, and it is now widely accepted that Tambora was the primary cause. The summer was so cold that crops failed across China, Europe and East Coast America, leading to unemployment, rioting and emigration. Thousands died of starvation and many more from dysentery and epidemics of typhoid, which were a direct result of the weakened population. The global impact of Tambora was so significant that some historians believe that it changed the world forever, both politically and economically.

Opposite **Lightning and molten lava bombs can be seen in the ash cloud during an eruption of the Eyjafjallajökull volcano on Iceland, an island that straddles the volcanically active boundary between two tectonic plates.**

Below **A very hot cloud of volcanic ash, dust and gases, known as a pyroclastic flow or glowing avalanche, cascades down the side of Mount Sinabung on the Indonesian island of Sumatra.**

Notwithstanding the terrible devastation caused by Tambora, there is an interesting footnote to the volcano's wider impact. The striking red skies seen in some of J.M.W. Turner's most famous paintings were the result of Tambora's volcanic dust, which hung in the air, scattering the sunlight. And the novelist Mary Shelley's experience of the wet, dark, brooding summer of 1816 inspired her creation of *Frankenstein*.

Of course, Tambora isn't the only example of a deadly volcano. Mount Vesuvius's legendary eruption buried the city of Pompeii, killing thousands. The energy released by Krakatoa's eruption was equivalent to 15,000 nuclear bombs and killed more than 35,000 people. In 1902, Mount Pelée in Martinique killed all but a couple of St Pierre's 30,000 residents. One of the two survivors was a prisoner in solitary confinement in the local jail; he was badly injured by the burning gases but, after recovering and being pardoned, he toured with the Barnum and Bailey Circus, becoming famous as the man who survived Mount Pelée. The 1985 eruption of Nevado del Ruiz in Colombia killed 20,000 and, although the huge eruption of Mount Pinatubo in the Philippines in 1991 *only* killed 772 people, it left more than 200,000 homeless. And the loss of life is certain to continue, with 350 million people – or 1 in 20 of us – living within danger range of an active volcano.

Eruptions aren't always deadly but there are other ways that volcanoes cause chaos in our modern world. In April 2010, for instance, the catchily named Eyjafjallajökull volcano in Iceland made big news, leaving newsroom anchors and reporters to struggle with the pronunciation for weeks. (For those interested, it's AY–uh–fyat–luh–YOE–kuutl, but try saying that quickly!) Nobody was killed by the eruption – rated only a 2 on the VEI – but it ejected huge quantities of fine, glass-rich ash into the jet stream, which resulted in around twenty countries closing their airspace to commercial aviation for a week, creating the highest level of air travel disruption since the Second World War.

While the destructive properties of volcanoes are well known, the frequency of eruptions is likely to surprise most people. Of the 1,500 currently active volcanoes on the surface of our planet (the vast majority, not included in this figure, are hidden away at the bottom of oceans), up to 30 erupt every year. And it's a sobering thought that volcanologists believe there's a 10 per cent chance of a Tambora-scale eruption happening in the next 50 years. Those are pretty decent odds of something catastrophic going off in our or our children's lifetimes. To make matters worse, scientists still have no way of accurately predicting exactly when a volcano is going to erupt. On top of that, history has shown that the most destructive eruptions come from volcanoes that aren't even on the radar of volcanologists, like the huge Pinatubo eruption in the Philippines.

Left **A nesting colony of great frigate birds is located next to the cinder cone of the dormant Bárcena volcano, on San Benedicto Island, in Mexico's Revillagigedo Archipelago Biosphere Reserve.**

ORIGINS OF LIFE

Below **Hot lava pours from the Pu'u Ō'ō volcanic cone, on the slopes of Hawaii's Kīlauea volcano, and travels down to the sea through lava tubes.**

Reading these stories, you could be forgiven for feeling that volcanoes, like mosquitoes, are one of those things the world would be better off without. But you'd be wrong (as you would be about mosquitoes). Yes, untold numbers of people – and animals – have had their lives prematurely ended by volcanic eruptions but here's the thing: literally none of us would be here without them.

Volcanism has been a feature of our planet for four billion years – almost the entire lifespan of the Earth. Volcanoes are responsible for our breathable atmosphere, the oceans, and the land above the sea. In short, life.

It's a bold claim, but take our atmosphere, which is unique in our solar system. Today it's made up principally of nitrogen and oxygen – 78 per cent and 21 per cent respectively. As it turns out, this is the perfect proportion. Clearly we wouldn't be able to breathe without oxygen, but if its level was much higher all kinds of strange and unwelcome things would happen, from giant insects to stuff catching fire more easily. Most worrying would be oxygen toxicity, which causes cells to break down.

Volcanoes are not responsible for the formation of oxygen – at least not directly – but they are totally responsible for all the nitrogen and were the original source of all carbon dioxide (CO_2), which, today, amounts to around 0.04 per cent of the atmosphere. We all know that CO_2 is a greenhouse gas, which is now threatening our planet's stability, but without it there would be no life since plants totally depend on this gas to grow. And what do plants produce when they photosynthesise? Oxygen. So, without volcanoes there would be no CO_2, and no CO_2 would mean no plants, no oxygen... and no breathable atmosphere.

The reason there would be no oceans without volcanoes is that almost all of the planet's water vapour came from molten rock and millions of years of intense eruptions. Once the Earth cooled, nearly 4.5 billion years ago, the water vapour condensed and formed the oceans.

Above **NASA** astronauts in the space shuttle *Endeavour* photographed the plume of erupting Kliuchevskoi volcano on Russia's Kamchatka Peninsula.

Opposite **Bwenge**, a young silverback mountain gorilla, surveys his surroundings from the crater rim of dormant Mount Visoke (also known as Mount Bisoke), on the border between Rwanda and the Democratic Republic of the Congo.

OUR PLANET'S MOLTEN INTERIOR

Picture a volcano and the chances are you'll imagine a large, cone-shaped mountain, possibly with smoke and fire bursting out of its pointy top. This kind of volcano is known as a stratovolcano or composite volcano because it's made up of layers of ash and lava. They can be powerful, violent and very dangerous. When they erupt, they often do so with huge force, spewing out lava and sending pyroclastic flows, sometimes containing rocks the size of trucks, down their sides. These are also the climate-changing volcanoes, like Pinatubo, whose ash cloud covered 2.6 million square kilometres in 30 hours. But stratovolcanoes are not the only kind of volcano.

Shield volcanoes get their name from their shape – an upturned warrior's shield. They may be very large, like Mauna Loa in Hawaii, but these types of volcanoes are not typically explosive. They can produce a lot of lava but, because they don't have steep sides like the stratovolcanoes, it flows more slowly.

Cinder cones are the most common type of volcano. They can grow quickly but are generally small. In behaviour and size they are rather like scaled-down versions of stratovolcanoes.

But the story of volcanoes starts with the planet's molten interior. To understand that, it's worth looking at the composition of the Earth. If you imagine it sliced down the middle, the cross section would show a series of concentric layers. At the very centre is a solid inner core – made almost entirely of iron – which has a radius of around 1,290 kilometres. Above that is the external core, 2,250 kilometres from top to bottom and mostly composed of iron. Unlike the inner core, this layer is liquid. And it's vitally important, as without it Earth would have no magnetic field to protect it against the sun's radiation. Stripped away by solar winds, the planet's atmosphere would leak into space and make our home as lifeless as Mars.

Moving up, the next layer is the semi-liquid mantle – composed of an inner and outer mantle – which, at a diameter of around 2,900 kilometres, represents 80 per cent of the Earth's mass or volume. This is the source of magma, or liquid rock, which, when it appears on the Earth's surface, becomes lava.

Finally, there's the crust – the bit we stand on and the oceanic seafloor. The Earth's crust is anywhere from 5 to 70 kilometres thick so, if you now imagine the Earth as an orange, the ratio of outer skin, or rind, to fruit flesh is not that much different from the planet's ratio of crust to mantle and core.

Given the depth of the Earth's crust, it's not surprising that magma from the partially melted, pressurised mantle is often able to break through surface weaknesses, particularly in oceanic plates where the crust is at its thinnest. It's also worth noting that the deepest we've drilled into the Earth is a very modest 13 kilometres, and it's unlikely we'll get a lot further. Jules Verne's famous story *Journey to the Centre of the Earth* would be impossible to complete – not least because you would need to deal with temperatures that rise to over 4,700°C. The temperature of fresh lava, by comparison – the hottest natural thing on the Earth's surface – is a quarter of that.

Opposite **Steam rises from the Ulawun volcano on New Britain, Papua New Guinea. Ulawun's last eruption was in August 2019.**

Overleaf **An enormous colony of chinstrap penguins gathers beside Mount Asphyxia (also known as Mount Curry) on Zavodovski Island, one of the South Sandwich Islands in the South Atlantic Ocean.**

ARCHITECTS OF THE PLANET

One way magma comes to the surface is at volcanic hot spots, which typically occur in the middle of continental or oceanic plates. Here, plumes of magma break through cracks in the Earth's crust. When that happens in the oceans, the outpouring of lava can lead to the formation of islands. Both the Galápagos and Hawaiian islands owe their existence to this kind of volcanic activity.

Kīlauea on Big Island, Hawaii, is one of the world's most active volcanoes. In 2018, lava flowed for four months without pause but it's been almost continuously active since 1983. In that time, Kīlauea's lava has covered 97 square kilometres of the island, destroying hundreds of houses in the process. In the last 35 years – and this is the important point – it's created nearly six and a half square kilometres of new land. That might not seem much, but volcanoes work on a different time scale, remaining active for tens of thousands of years.

Big Island is about a million years old, though still the youngest of the Hawaiian Islands, and during that time its collection of five volcanoes (two, including Kīlauea, are active, and one, Mauna Kea, is the tallest mountain in the world when measured from its base on the seafloor) has created over 6,440 square kilometres of land.

Islands are great examples of the land-building power of volcanoes, but it's not just islands: nearly every square metre of our planet has its origins in Earth's molten core. And it's a process that never stops.

As soon as lava has cooled and hardened, it becomes a platform for life. Plants are generally the first to get a foothold. On islands these could arrive on the wind, or washed in on currents. They've even been brought in accidentally on the feet or feathers of seabirds. Once a new island has established a successful plant community, it provides opportunities for animals. The speed of succession, however, depends on how far the island is from the nearest large landmass. On Hawaii, the world's most remote archipelago, it's thought that it took over 30,000 years for each new species to become established. That seems an unfathomable length of time to us, but in geological time it is still the blink of an eye. The Galápagos Islands – probably the most famous group of islands in the world – were millions of years in the making, popping up on a still active hot spot, slowly moving on a geological conveyor belt. Like Hawaii, they are isolated from continental landmasses but, nevertheless, are extraordinarily rich in wildlife.

Right **A washed-up coconut palm sprouts on a stony beach on the south shore of Tutuila, American Samoa.**

Overleaf **Red-hot lava from Hawaii's Kīlauea volcano flows directly into the sea, creating new land.**

THE DRAGONS OF GALÁPAGOS

On islands with still active volcanoes, life can certainly be precarious, but on Fernandina Island, in the Galápagos, one animal has turned these dangers into an advantage.

Every year, in May, around 2,000 female land iguanas, all heavy with eggs, make an arduous, two-week trek from the coast to the top of La Cumbre, Fernandina's active volcano. Their goal is to lay their eggs in the warm volcanic ash, which is the perfect temperature for incubation. The first to arrive nest on the crater rim but the options here are limited. So, for many females, the only choice is to continue to the bottom of the crater. Space is less of an issue but it's a perilous journey. There are no clear paths down – any made in previous years would have been lost to a combination of earthquakes, landslides and lava flows.

La Cumbre's crater is over 800 metres deep, but each female iguana will have to travel many times further as she picks her way down the steep, rocky slopes of the volcano. On the more precipitous sections, it's easy to dislodge rocks, which can quickly turn into avalanches. It's why, for each iguana, the journey's biggest danger comes from other females descending along the same route. Land iguanas are tough, bulky animals – growing to a length of one and a half metres – but there's no defence against a large rock or boulder speeding downhill towards them. For some, the trip down will be their last, and they will be buried under a pile of rocks.

Right **A female land iguana pauses on the volcanic rim of La Cumbre on Fernandina Island, Galápagos, before making her descent to the bottom of the crater where she will lay her eggs – a journey that could take several days.**

Below **A female land iguana on the slopes of La Cumbre, a shield volcano on Fernandina Island. In 1968, a major eruption caused the caldera floor to drop by around 300 metres – a force that was felt on the other side of the world. The lake you can see in the background reformed in the 1970s.**

Above **A female land iguana digging her nest in the volcanic ash on the crater floor of La Cumbre, Fernandina Island, Galápagos. The ash is the perfect temperature for incubating her eggs.**

Opposite **A female land iguana. The population of land iguanas on Fernandina Island is estimated at between two and three thousand, though no one knows for sure.**

For those that succeed, it will have been an eight-hour trek to the bottom, and there are still other challenges to come. Space may not be as limited as on the rim but there's still competition for the best sites and, once they've started to dig their nests, female iguanas are extremely protective of their territory. A vigorous head bob means keep out, stay away. Ignored warnings are met with force – and female land iguanas can be very aggressive. Fights can be vicious.

When the right spot has been found, a female will spend several days digging her nest, after which she will lay up to 20 eggs inside before covering up the entrance. This will be her final act of motherhood as it'll be the volcano's warm embrace that will incubate the next generation of land iguanas. She may stick around for a few days to guard her nest site from others but there's no dodging the final challenge – the journey back to the top, which is no less dangerous than the one down. One thing's for sure: no female would go through such an epic trial if the end result wasn't worth it – proving the importance of the incubating power of La Cumbre.

INTO THE CALDERA

If descending La Cumbre is hugely hazardous for land iguanas, it can seem almost suicidal for humans. We're obviously much heavier than iguanas so it's even easier to set off landslides and avalanches. That's one reason why more people have been in space than at the bottom of Fernandina's volcano. But if you want to film the extraordinary breeding behaviour of this island's land iguanas you must follow them down – and that is no mean feat.

Fernandina is one of the most pristine islands in the Galápagos. It's also remote, uninhabited and hostile – with temperatures regularly exceeding 45°C. Any film shoot on the island requires months of planning and the close cooperation of the National Park. There is no fresh drinking water on either the volcano's desert rim or in the crater so, for our 15-day shoot, it all had to be brought in on boats and transported up by porters – a perilous two-day trek for each load. The ferrying of supplies was made more difficult by the strict park rules on how many people are allowed on the island at any one time – restrictions that limited the number of water trips that could be made. So Toby Nowlan, the field director, had to work out precisely how much water each of the six crew would need over the period. As Toby said, 'It meant that water had to be rationed almost to the last drop.' And, of course, it wasn't just water: food and camping and filming equipment also had to be carried up the volcano and into the crater. It's fair to say that few shoots require such meticulous, military-style planning.

But even the most detailed preparations can't account for every eventuality. When it comes to volcanoes, some things must inevitably be left

Below The film crew's camp on the crater rim of La Cumbre, Fernandina Island, Galápagos. Several days after they pitched their tents, a drone shot of the crater revealed that the campsite was circled by a radial fissure – a fault probably caused by an earlier earthquake. Some time in the future, this chunk of crater wall will almost certainly drop to the bottom of the caldera.

in the hands of the gods. A stark reminder of this occurred just a week before the crew arrived, when La Cumbre erupted, spewing lava down one side of the volcano. Fortunately, it wasn't on the side we planned to work on so, after careful consideration, the shoot went ahead.

The only way of getting to the island is by boat and, after a tricky landing of people and supplies, the next step was carrying everything up to the rim: a ten-hour journey across and up a lot of very hostile terrain. Before setting off for the top, cameraman and Galápagos expert Richard Wollocombe looked up to the cloud-covered rim and wryly observed to the rest of the team, 'You've got to be totally crazy to go up there.' Richard had once made the backbreaking trek to the rim but never into the crater itself.

Once at the top, the crew, physically exhausted by the uphill march with kit and supplies, pitched our tents on the edge of the crater rim – ten metres from the drop-off. It was a distance everyone thought was sufficient until a few days later we used a drone to get a shot of the camp. Only then was the campsite's seemingly precarious nature revealed. The bit of rim the team had chosen to pitch our tents on was, in fact, a huge wedge of crater wall that looked as if it could give way and plummet into the abyss at any moment. Of course, given geological timescales, it could have been like this for a very long time so, thinking rationally, we reminded ourselves that the chance of something catastrophic happening in the week we were to spend on the top was extremely low. But other things still played on our minds.

Earthquakes are common after an eruption, and one measuring 9.0 on the Richter scale had been recorded during the eruption that had occurred shortly before our arrival. Could further aftershocks cause a sudden reshaping of the rim we were camped on? It was a possibility, even without an earthquake.

Every night, as the ground cooled and contracted after the intense heat of the day, the team would be woken by a terrifying grumble from rock that had come loose and thundered down the crater slopes. As Richard said about one particularly large landslide, 'I thought we were going to slide into the abyss, and I rushed out of the tent fully expecting to be engulfed in chaos. After I realised we were safe, I walked to the edge of the crater to see if I could make out where the landslide had taken place. In the light of the full moon, I could see that so much rock had fallen that a huge cloud of dust was still billowing above the crater floor and over the very spot where we hoped to be camping in a few days' time.'

As hairy as camping on the rim felt, the team knew the biggest danger was the journey to the bottom – just as it is for the iguanas. The first task was finding a route down. For that, the crew looked to Tui De Roy. Tui is a world-renowned wildlife photographer, naturalist and writer who grew up in the Galápagos – and a tougher 66-year-old would be hard to find. She was the first to discover and document the iguanas' descent of La Cumbre and their nesting in its crater, and is one of a very small number of people who have been to the bottom. The route she'd taken previously had long since gone but after much searching a path was picked out – though no one could really tell whether it was safe or not.

Below Cameraman Sam Stewart films a land iguana as she makes her way down to the crater floor. Sam's helmet is a protection against falling rocks – a common occurrence in La Cumbre.

Above **Camping on the edge of the La Cumbre volcano on Fernandina, 800 metres above the crater floor. It wasn't until the crew started filming with the drone that they realised there was a large crack in the earth running around their campsite. It's only a matter of time before this section of crater wall falls into the chasm.**

Thinking about another sudden eruption while moving down slopes of loose lava that sink away underfoot is bad enough, let alone the very real risk of being caught in an avalanche of boulders, particularly when walking underneath towers of precariously balanced rock. So, trying to focus on the purpose of the trip – filming the descending iguanas – wasn't exactly easy when gripped with a fear of several tonnes of rock barrelling towards you.

Thankfully, the crew all reached the bottom without any injury: a huge relief for all. Everyone naturally assumed the most dangerous part of the expedition was behind us. But then Volcan Chico on the neighbouring island of Isabela erupted. Not long after, a change in the wind direction brought the eruption's foul-smelling toxic gas to Fernandina. The blue gas cloud sank into the crater and soon it was so thick we could barely see the tents around us. Despite all the planning, this wasn't something we'd prepared for! Breathing through wet cloth to prevent the gas condensing into sulphuric acid in our lungs, we realised we were trapped. No helicopter could even attempt a rescue in such visibility and trying to climb out of the crater in the dark would have been suicide. So we had no option but to wait and hope that the winds changed. Salvation came at 3 a.m., when a change in wind direction brought fresh air. As Toby said the following morning, 'Rehydrated scrambled eggs at breakfast never tasted so good!'

TALE OF A VAMPIRE

Volcanoes are also a driving force behind the planet's diversity as, over time, new land provides the perfect platform for the evolution of new species. The Galápagos Islands are a legendary example. Here, through a process known as adaptive radiation, one ancestral species of finch, for instance, has diversified into 13 new forms.

The latest addition to Galápagos's finch list lives on the tiny, uninhabited and remote island of Wolf (an extremely difficult place for a small bird to have got to – but not easy for a film crew either as it required a 12-hour boat journey from Santa Cruz, Galápagos's principal island). Nobody knows when the first finch arrived on Wolf but the castaways (thought to be sharp-billed ground finches blown in from other Galápagos Islands to the south) undoubtedly faced unfamiliar challenges on their new home – an area of less than 0.8 square kilometres and the remains of an extinct volcano.

Finches typically feed on seeds, insects and nectar, and all these are certainly available on Wolf – but not in the quantities required. So, to survive on Wolf, the finches had to adapt; they needed to find another source of food. And Wolf's temporary residents, the Nazca boobies that commute daily between the island and their feeding grounds out to sea, provided the perfect solution.

Below **Pairs of Nazca boobies on Wolf Island, Galápagos. After a day spent fishing out in the open ocean, the birds return to the island and renew their bonds with some beak clacking and mutual preening.**

Above A vampire finch drinks the blood of a Nazca booby. The finch gets the blood to flow by cutting one of the seabird's large flight feathers. This very un-finch-like behaviour is an adaptation to the island's scarcity of food and water. Strangely the boobies don't seem to mind.

At the right time of the day, it doesn't take long to witness what is surely one of the most bizarre behavioural traits of any bird – especially a bird that looks as innocent and mild mannered as our house sparrows. To supplement their diet, amazingly, Wolf's finches drink blood. One will jump onto the long tail feathers of a booby and from there will peck at the base of the seabird's larger flight feathers until a wound opens up and blood starts to flow. The finch then laps up the drops or draws its beak through the blood-soaked feathers. It's why these birds are known as *vampire finches* and now scientifically acknowledged as a unique species.

Surprisingly, most boobies don't seem to mind having a blood-drinking vampire on their tails, which raises the question, why not? One theory is that the vampirism evolved from something that might well have benefited the boobies, like parasite removal. So, presumably, the seabirds still believe the little vampires are providing a service.

LIFE FROM DEATH

Volcanic islands don't last forever. When they cease to be active, wind and rain slowly wears away the rock, as it does everywhere on Earth. But much more significant here is the weight of the land, which causes these islands to sink back into the seafloor. That would normally be the end but not always. In the tropics, most islands are circled by coral reefs, which depend on sunlight to grow. So, as the island sinks, the reef builds up in the shallows. Eventually, all that's left of the original volcanic island is a lagoon surrounded by a fringing reef. It's become an atoll.

All atolls owe their existence to volcanoes. Currently, there are over 400 atolls across the tropical oceans and they come in many different shapes and sizes. One of the largest, at 160 square kilometres, is Aldabra in the Indian Ocean. It's thought the island emerged around 20 million years ago, becoming an atoll just 200,000 years ago.

Aldabra is now politically part of the Seychelles more than 1,190 kilometres away. To get there by air, you have to charter a plane to Assumption Island, which is extremely expensive, and then it's an hour's speedboat trip to Aldabra. The alternative is an arduous four- or five-day boat journey, so the only reason to do that is cost. This remoteness is what makes Aldabra one of the most inaccessible atolls on the planet – and, after nearly 50 years of protection, one of the most pristine. Sir David Attenborough has called it 'one of the wonders of the world'.

Except for a small scientific base on one corner, Aldabra is uninhabited. Today, it's a vital refuge for some unique species, such as the Aldabra rail, the only remaining species of flightless bird in the Indian Ocean (apparently an example of iterative evolution – it's the second time a flightless rail has appeared on the island, after the first one went extinct 136,000 years ago), and the last of the region's giant tortoises.

Aldabra giant tortoises were also once on the brink of extinction but there are now 100,000 living on the atoll. Wherever you go on Aldabra there are giant tortoises: in the pine forests, the mangroves, on the raised and pitted limestone, known as *champignon*, under the casuarina trees along the coast and sometimes on the golden sandy beaches by the edge of the sea. The largest tortoises weigh about 250 kilograms. The oldest – in excess of 200 years – would even have been wandering the atoll when, in 1874, Charles Darwin and other prominent naturalists wrote a letter to the Governor of Mauritius appealing for the protection of these incredible reptiles. The naturalists were especially concerned about plans to exploit the islands' mangrove trees. As the letter said, 'If this project be carried out, or if otherwise the island be occupied, it is to be feared, nay certain, that all the Tortoises remaining in this limited area will be destroyed by the workmen employed.' Fortunately, the Governor took note and the plans were shelved.

Giant tortoises can go without food and water for many months at a time, which makes them ideally suited to Aldabra, where there's a regular scarcity

Opposite top **At 160 square kilometres, Aldabra in the Indian Ocean is one of the world's largest atolls. Uninhabited – except for a small research base – Aldabra is also one of the planet's most remote islands.**

Opposite bottom **Giant tortoises heading for the shade of a coral cave. Temperatures in Aldabra can rise to 40°C – far too hot for giant tortoises – so, every morning at around 9 a.m., during the dry season, these lumbering giants head for cover.**

of both. At certain times of the year, it's not unusual to see giant tortoises grazing on dead leaves. But there's one thing their bodies are not equipped to deal with, and that's the heat of the tropical sun. Daytime temperatures on Aldabra can rise to over 40°C – way beyond a tortoise's optimum temperature. So, at around 9 a.m. each morning, giant tortoises 'race' for the shade of trees, clustering together in groups till about 4 p.m. On the eastern side of the island, in an area called Grand Terre, the options for shade are scarce so there's competition for the best spots. A small limestone cave has room for about 60 and the top shade tree in the area has space for just over 200... at a push. The last to arrive may try and clamber onto the backs of others or risk death in the sun. The bleached carapaces scattered over the island's limestone surface are evidence that some just wander too far from the shade or arrive at their chosen cover too late.

On Grand Terre, when the tortoises emerge from cover in the afternoon, they spread out over the tortoise turf, a wide strip of cropped grass, which begins a few metres from the sea. While grazing, they step over and around washed up beach litter. Despite Aldabra's remoteness, the prevailing currents dump huge amounts of flotsam and jetsam onto the land. Surprisingly perhaps, the most common items are flip flops but there's everything from fishing nets and buoys to plastic bottles and containers, printer cartridges, toothbrushes and plastic toys. The sight of a giant tortoise on the world's most pristine and protected atoll grazing around a sea of plastic debris is a poignant reminder not just of the importance of waste management and recycling but also of the state of our planet. And this isn't the only man-made issue Aldabra needs to worry about.

The biggest danger to the tortoises – and indeed, the whole of Aldabra – comes from our warming planet. Aldabra, like most atolls, sits only two or three metres above sea level. Being so low lying means it's particularly vulnerable to the effects of climate change, which is melting ice sheets and causing the oceans to rise. Ironically, even without humans, the rise in sea level could come from volcanoes themselves, as they are a natural source of carbon dioxide. So volcanoes may give life to atolls but they can also take it away. Indeed, Aldabra has been completely inundated by the ocean at least once in its life – resulting in the loss of all its flora and fauna.

Right **An Aldabra giant tortoise. Once on the brink of extinction, there are now around 100,000 living on this atoll. These huge reptiles can live for over 200 years and weigh as much as 250 kilograms.**

SHIFTING CONTINENTS

The power of the planet's molten interior has a much bigger impact on our incredible diversity of species through another more powerful force: a process known as plate tectonics or continental drift.

The planet's surface is made up of different plates – pieces of the Earth's crust and upper mantle, known as the lithosphere. They are a combination of continental and oceanic plates, which fit together like a jigsaw puzzle (though, if it was a jigsaw and each piece was a plate, it wouldn't be much of a challenge on a long wet weekend as there would be only 17 or so pieces: seven big ones representing the continents and the rest smaller pieces, the minor plates). They range in size from the Pacific Plate, the biggest at over 103 million square kilometres, to the Burma Plate at 1.1 million square kilometres.

Above **Molten lava and huge quantities of ash spew from the erupting Tolbachik volcano on Russia's Kamchatka Peninsula.**

For over three billion years, through plate tectonics, the various pieces of the lithosphere have been on the move, shaping and reshaping our continents. Put simply, magma is heated in the mantle, causing it to rise to the surface and, as these convection currents spread out sideways, it pulls the plates apart. It's a very slow process – on average, the plates move three to five centimetres a year – but over geological time the results are spectacular.

Let's take the last 250 million years, and turn the events into an imaginary time-lapse. At the start, you would see the planet's landmasses joined in one supercontinent known as Pangaea (deriving from the Greek meaning 'entire Earth'). Over the next 20 million years – or a few seconds in our time-lapse – this supercontinent starts to break into two huge landmasses: Laurasia in the north, which includes the future Europe, Asia and North America, and Gondwana to the south, the precursor to the continents of Africa, South America, Antarctica and Australasia, and the landmass of India.

As the next 135 million years speed by, these two megacontinents split into seven large landmasses – now recognisable as today's continents – plus India, which, at this stage, was doing its own thing, independent of the other landmasses.

The last stage of our time-lapse, covering the next 100 million years of Earth's history, shows the continuing journeys of the great landmasses to, roughly, the present day: North America heads west, South America northwest, Antarctica goes south, Australasia moves east, Europe and Asia rotate clockwise, Africa floats north east, and India keeps going north until it crashes into Asia (creating the Himalayas in the process).

By 20 million years ago the world looked much as it does now – the only noticeable difference being that North and South America had yet to join, and Britain was still physically connected to Europe.

The reason plate tectonics have made such a huge difference to the planet's diversity of species is simple. On a stable supercontinent like Pangaea, there was no pressure for species to adapt and change, so biological diversity essentially flat-lined. But on slowly separating and isolated landmasses everything was in flux – new habitats and climates began appearing, and these changing environments forced each new continent's cargo of species to adapt and evolve. In other words, evolution and diversity went into overdrive.

Unsurprisingly, it's at plate boundaries that volcanic activity – leading to eruptions and earthquakes – is at its strongest. Indeed, it's where over 80 per cent of lava is emitted. Not all the plates are being pulled apart, however. Some are converging in what's known as subduction zones. Where this happens, a lighter oceanic plate is pulled under thicker continental crust. The best example of this is around the Pacific Plate – the most volcanically active region on Earth, known as the Ring of Fire. This horseshoe-shaped ring spans a distance of 40,234 kilometres and has more than 75 per cent of the world's active and dormant volcanoes. It's why, for instance, the Kamchatka Peninsula, on the edge of the Asian and Pacific plates, has so many volcanoes.

Active volcanoes have the power to reshape the planet, but even those long dormant can have an impact at the surface.

A SLEEPING GIANT

Yellowstone National Park is home to the world's most famous dormant volcano – or supervolcano, on account of an eruption 650,000 years ago, that volcanologists have labelled an 8 on the VEI. Nobody knows when it will erupt again but it's far from dead. Here, groundwater – percolating down through cracks in the Earth's crust – is heated by magma. When it reaches boiling point the pressure forces the water up to the surface, bursting out as geysers. This cycle of boiling and bursting can be very reliable. The aptly named 'Old Faithful' – Yellowstone's legendary geyser – erupts to a height of around 50 metres every hour or two, 24/7.

In addition to geysers, Yellowstone has literally thousands of other thermal features, including fumaroles (openings in a volcano which release gases), hot springs and pools of bubbling mud, making it the largest geothermal area in the world. And all this underground heat is particularly important to one of Yellowstone's residents: river otters.

Otters, part of the mustelid family, live life in the fast lane. They have a high metabolic rate, which means they need a constant supply of food – and that requires rivers to fish in. But Yellowstone is one of the coldest places in America, with temperatures dropping to as low as minus 30°C: cold enough to freeze even the fastest-flowing water. That would be the case in Yellowstone, too, were it not for its thermal springs which keep some rivers ice free even during the coldest winters. It means the otters can fish all year round.

Right **A North American otter eating a fish it has caught in a partly frozen lake in Yellowstone National Park, Wyoming.**

Below **The cone of geyserite sinter deposits formed first in 1022 at Castle Geyser in Yellowstone National Park. The geyser erupts every 10–12 hours for about 20 minutes.**

When winter begins to bite, the otters' fish-catching skills sometimes attract the attention of other Yellowstone residents. Coyotes are the ultimate opportunists and they know that a thermal river often means otters, and otters mean fish. Otters like to eat their catch on the banks so for a quick-witted coyote there's a chance to steal the mustelids' prize – though success isn't guaranteed. Clearly the coyote we filmed during our shoot was less wily than required because, despite numerous attempts, it was never able to get the better of the otters.

There are only a thousand known geysers around the world, and many are found in Kamchatka's Valley of the Geysers. In May, brown bears, fresh out of hibernation, make the most of the Valley's warm, snow-free ground to graze on grass fertilised by mineral-rich spray from the valley's geysers. At this time of the year, it's their only option for a meal as – even in May – the rest of Kamchatka is still deep in snow.

Above **In May, when the rest of Kamchatka is still covered in snow, the mineral-rich grass of the Valley of the Geysers is much needed forage for brown bears, fresh out of hibernation.**

Each morning, when steam from dozens of geysers fills the small valley, as many as a dozen bears can be seen grazing on the steep slopes and valley bottom. Interactions are kept to a minimum as clearly nothing must get in the way of harvesting the first grass of the year. (Fortunately, the bears weren't interested in interacting with the film crew either, though park rules dictate that we be accompanied by an armed ranger at all times, just in case.) The bears seem equally oblivious to the dangers from all the boiling water that spurts, bubbles and flows around the valley. With so much scalding water around, one could almost believe the bears had heatproof paws as they move with a lumbering ease around the valley. Nevertheless, the hazards are very real, as cameraman Rolf Steinmann found out.

A STEP IN THE WRONG DIRECTION

After a week of carefully negotiating the Valley's densely packed thermal features while carrying heavy filming equipment, Rolf accidentally stepped back into a small patch of ground soggy with boiling water. He felt a sudden burning around the ankle of his left foot, just above the cuff of his boot. By the time he'd pulled his shoe and sock off, the skin on his lower shin had started to peel off. It looked serious. A speedy evacuation was clearly needed, though that is easier said than done in this remote part of Kamchatka.

The Valley of the Geysers is only accessible by helicopter, which would normally take time to arrange. Happily, however, after the bad luck of the injury there was some good fortune – a helicopter was already en route to the Valley with a group of German sightseers, the first tourists we'd seen since arriving. So, less than a couple of hours after the injury occurred, Rolf was in a helicopter heading back to town, to be met at the airport by our local fixer, who took him straight to hospital.

Naturally, I assumed this was the end of the shoot and, while Rolf was making his way to the hospital, I packed up the equipment, expecting to fly out the following morning. Later that afternoon, though, I got a message from the fixer saying things weren't as bad as originally thought and, after treatment at the hospital, Rolf was already on his way back to the Valley of the Geysers.

I got the full story when Rolf returned. The doctor had told him that it was 'only' a first-degree burn which required keeping clean with daily dressing changes. Antibiotics and painkillers would take care of the rest. With this prognosis, Rolf was happy to continue with the shoot. Those are the words

Above **Cameraman Rolf Steinmann** threads his way through the scalding pools in Kamchatka's Valley of the Geysers. A misstep, later in the shoot, resulted in a very nasty burn on his shin which required evacuation and immediate hospital treatment.

every director wants to hear on location but they were certainly unexpected – even from Rolf, who has a legendary reputation for a tough, uncomplaining approach to filmmaking no matter how extreme the conditions of the location.

By the following day, it was clear compromises were needed. As much as Rolf wanted to carry his own camera and tripod, this wasn't possible with his injury. Neither was sticking to the same filming routine we'd had before the accident. Even with the painkillers it was obvious Rolf's ankle was troubling him so we cut back on the ambition – moving around much less and staying out for shorter periods.

We changed the dressings on Rolf's burn as often as recommended and over the following days there did seem to be a small improvement in the look of his injury, but Rolf was still in pain. He soldiered on and got a few very nice shots, which made the final cut, but by the time the shoot ended a couple of days later it was clear that Rolf needed a second opinion on his shin. Back in Germany, he saw a burns specialist and was told that no, it wasn't a first-degree burn, but a third-degree one, the most serious kind. An immediate skin graft was needed to prevent any further complications. In the end, Rolf spent four weeks in hospital. We'd been due to film in a lava lake in Vanuatu a week after returning from Kamchatka, which obviously had to be postponed. I thought Rolf might never want to be anywhere near a volcanically active area again, but 18 months later we were on Tanna Island, in Vanuatu, filming the active Yasur volcano and its impressive lava lake. The best wildlife cameramen are a very determined lot!

MOTHER EARTH

One of the major benefits of volcanoes is their role in fertilising the planet. Ash clouds may be inconvenient if they happen to be blowing along commercial aviation routes, but the very largest can contain a billion tonnes of minerals, like iron, magnesium and potassium, which are recycled from deep in the Earth's core. That's why areas with the most volcanic activity are also the most fertile. Kurile Lake, in Kamchatka, is one such beneficiary.

The lake sits in the crater of a dormant volcano – the result of one of the world's most powerful eruptions. But active volcanoes still ring its shores and mineral ash from regular eruptions has made the waters so fertile that Kurile Lake now supports the greatest diversity of salmon species on Earth, including chum, pink and sockeye – six million of them. The sockeye are the most numerous, and Kurile Lake has more than anywhere else in Asia.

Nutrients from the lake's active volcanoes make the water rich in plankton, which is what the salmon feed on. After eruptions, the population of salmon explodes. The fish spawn in the lake, and the activity attracts bears. Brown bears are mostly solitary but for a few weeks in summer they put aside their differences and, in Kurile Lake, gather in the greatest density found anywhere on Earth.

Below Cameraman Tom Walker films fishing bears with a gyro-stabilised camera system, Kurile Lake, Kamchatka.

There may be plenty of salmon to go around, but catching them early in the season is not easy. Initially, there's a lot of running and splashing, but little catching. With so many fish on show it's clearly hard for a bear to hold back, particularly inexperienced bears. Early in the season, however, the oldest, wisest bears take a different approach. They swim out, away from the shoaling salmon, and into the middle of the lake where they go diving for dead fish. In some parts of the world, it's said that brown bears don't like getting their ears wet, but – even if that is sometimes the case – it's not true in Kurile. Here, the old-timers go diving for fish carcasses. The rewards may not look much, but harvesting the remains off the bottom of the lake doesn't use many calories and they can take their time. The diving bears use two different techniques to get to the bottom: either they duck-dive, kicking their legs into the air, while pulling themselves underwater with their paws; or they use the 'pencil technique', where they drop feet first, and their head is the last part of their body to go underwater. It's hard to say whether one is better than the other.

When the salmon are focused on spawning, thousands begin to power their way up the network of streams. Then the bears go into action, diving on them from all angles. Fishing is now so easy that even those with no previous experience, like this year's cubs, get in on the act. There's so much choice and opportunity that, for some bears, it's hard to know which salmon to chase. And even then, when one is landed, it's hard to ignore others that come within striking distance. Often, fish are left flapping and gasping on the sand, while the bear that caught them charges off to chase others.

For the crew, having large bears running past them – sometimes within touching distance – certainly quickened the pulse. But the bears were so absorbed by the salmon that they barely gave the team a second thought, which put everyone at ease. Except once perhaps… As Toby, the field director, explained, 'Hearing a thundering gallop, we turned to see two young males charging towards us at full speed. Nicolai, the park guard, leapt up and in seconds had let off a cloud of bear spray in front of the pair. The young bears stopped short at the cloud, sniffing it in disgust, and sauntered off.' As it turned out, this wasn't a threatening act – the bears were just chasing each other around and had been so wrapped up in their squabbles that they hadn't seen the crew.

Kurile Lake's salmon run is a real-life bear's picnic, and one so laden with food that each bear can afford to be fussy, and seemingly wasteful. This event is all about calories and getting as many as possible – so why bother with the lower calorie flesh when you can feast on salmon eggs – caviar – that are 40 calories an egg? You could say it pays the bears to be decadent. It's why, during the salmon run, each bear manages to consume 100,000 calories a day – 50 times more than the average adult human.

And so there's only one conclusion here – by supporting such huge concentrations of fish, Kamchatka's volcanoes have made life surprisingly easy for its bears.

It's not just Kamchatka, though. Our entire planet depends on minerals brought up by volcanoes from Earth's molten interior. Tanzania's Serengeti, in East Africa's Great Rift, would certainly be very different without these underground forces. Mineral ash from Tanzania's active volcanoes, like Ol Doinyo Lengai, has resulted in grass so rich that it supports the greatest gatherings of animals on the planet. Every year more than a million wildebeest migrate to one corner of these great grasslands to have their calves. The grass here is rich in calcium and phosphorous – essential minerals for pregnant females. Here, for around three weeks, they give birth at a rate of 12,000 a day – that's 500 an hour. Of course, with this much meat on the hoof there are also unrivalled concentrations of predators – lions, hyenas, wild dogs, leopards and cheetahs. Each hunter has a different strategy but the aim is the same: to make the most of the short explosion of wildebeest calves. This hugely productive ecosystem has been around for nearly two million years but it wouldn't have been possible without the regular recycling of minerals from the Earth's core.

Opposite top Duck-diving brown bear. Early in the salmon spawning season, when the fish are generally too quick to catch, the more experienced bears know that there are easy pickings to be had on the lake bed, where the bodies of dead salmon have drifted.

Opposite bottom Up to six million salmon spawn in Kurile Lake each year and it attracts the greatest density of brown bears in the world. In a single day, a bear can consume 100,000 calories – 50 times more than the average person.

Below Mother and cub play on the edge of Kurile Lake – behind rises the Ilyinsky volcano. Mineral ash from this and other nearby volcanoes makes the lake's waters extremely fertile. It's the reason why Kurile Lake has such a huge population of salmon.

VOLCANO BIRDS

The effects of volcanic activity create enormous challenges for anything living in the vicinity – hardly surprising when you think about pyroclastic and lava flows, ash deposits, earthquakes and so on – but in Africa there is one animal that has come to depend on an active volcano for its very existence.

Tanzania's Ol Doinyo Lengai is Africa's most active volcano – and part of the East Africa Rift: a geological fault running for 5,955 kilometres along the African Plate from Lebanon to Mozambique. On Lengai's northern flank it has given rise to Lake Natron, which is one of the world's most corrosive bodies of water. At the base of the volcano, a concentration of chemicals seeps up from underground springs – making the water so caustic it can burn skin. On top of that, water temperatures in the lake can exceed 60°C. Across our entire planet, you would struggle to find a more unpromising place for life. At least, that's what you'd expect.

Nearly two million lesser flamingos live in East Africa, and all depend on this toxic volcanic lake to breed. But it isn't quite as simple as that. They can only breed on the lake when the water levels are the perfect depth – low enough for the flamingos to lay eggs on their raised mounds out in the middle of the lake. The perfect conditions don't happen very often. In fact, they can go five years or more without ever breeding successfully so it's just as well they are long-lived birds (in the wild they have a life expectancy of between 20 and 30 years).

The flamingos will attempt to breed every time the water levels look as if they're heading in the right direction but more often than not the lake will suddenly flood, either from rain or from the many streams running into it, and the flamingos will be forced to abandon the attempt. This unpredictability is probably why the flamingos seem to have no specific breeding season, with nesting events documented in every month of the year – but then nesting data is surprisingly limited for a large animal that gathers in vast numbers. The reason for this is the size and remoteness of Lake Natron, which can't be traversed on foot or by ordinary vehicles. The lake is so large that it's impossible to see the breeding colony out in the middle from the edge, so the only way of knowing whether the birds are breeding or not is to make regular overflights – an expensive undertaking. But when conditions are right, lesser flamingos come from all over East Africa to breed.

How they know when the water depth is perfect for nesting no one knows, but some come from hundreds or even thousands of miles away. Arriving at Natron, the flamingos fly in past Lengai and low over the lake in groups, often in V formation. They head towards the centre of the 950-square-kilometre

Top left **Two million lesser flamingos live in East Africa and all of them nest at Lake Natron, Tanzania. But they can only do this when the lake's level drops enough to expose islands of salt at its centre. In the background is Ol Doinyo Lengai, Tanzania's most active volcano. A concentration of chemicals from Lengai has made Lake Natron one of the world's most corrosive bodies of water.**

Bottom left **Before their flight feathers have developed, lesser flamingo chicks can gather in crèches of thousands.**

lake, where rafts of soda crust have been pushed by winds into islands. It is here that the flamingos establish their nesting colony. There's good reason for this – out in the middle they are several miles from shore and, thanks to the ring of caustic water, well out of range of land predators like jackals and hyenas. Adult lesser flamingos have no defence against Africa's predators except flight. The chicks, flightless for the first three months, would be even easier pickings without the lake's protection.

In a good year, up to half a million pairs nest on Lake Natron. Each pair builds a raised mound, made of brine and salts, that hardens like concrete, where a single egg is laid. The mound is a protection against a sudden rise in water level but the top is also slightly cooler than the surface temperature. After around 30 days of incubation, the chick hatches. It's a harsh initiation; even on the little mound, temperatures can rise to 54°C. So, during the hottest time of the day, the chick may shelter under its parent's wings.

As soon as it is born, the chick is fed a rich liquid made up of the lake's algae and traces of the parent's blood. A week later, it's strong enough to leave the mound and join up with other chicks, forming an ever-growing crèche. Since flamingos are asynchronous nesters, the crèches contain chicks of quite varying sizes.

Each day, the adults leave the colony and head out to feed. The volcanic lake not only provides a safe nesting site but also good supplies of their favourite food – blue green algae, or Spirulina, which blooms in the salty water on the fringes of the lake. The algae is a highly productive food source and the flamingos are one of the few animals able to exploit it, which they do with a bill containing thousands of hairy structures – filtering the water with sideways sweeps of their upturned bills. It's this algae that gives the flamingos their pink colouring – even turning their eyes pink.

Above **Lesser flamingos on the move. Ten or so days after hatching, the chicks make an epic trek from the centre of the lake to freshwater springs on the lake's margins. The journey has many challenges, not least predatory marabou storks which pick off any weak or injured chicks.**

As the middle of the lake dries out, the chicks are driven by a need for fresh water. Leaving the nesting colony, they embark on a journey of nearly six miles, which they do without any parental support. Since none can yet fly, this trek must be done on foot. Thousands of chicks – of all ages and sizes – set off on an epic march over razor-sharp soda surfaces and sticky caustic mud. Not all will make it. Exhaustion will take its toll on many – particularly the smaller chicks who don't have the size and strength to pull themselves through sections of sucking mud. Others will get weighed down with anklets of hardened soda, which can become so cumbersome that they're unable to walk more than a few paces at a time. Failure to keep up with the huge moving crèche is invariably fatal.

And then there are the marabou storks. They have the legs of a flamingo but beaks like Roman swords. Sometimes known as the undertakers, a more sinister-looking bird would be hard to find. Unfortunately for the chicks, the caustic soda that rings the lake is no barrier to these predators, and just their presence causes panic amongst the moving columns of chicks. It's exactly what the marabous want. To avoid being trampled by the older chicks, the smallest stick to the outside and that brings them in range of the storks. Any that stray away from the column are even easier targets. The chicks' only line of defence is to run but that's not much good when you're being pursued by an animal with legs several times longer. When a chick is caught it is swallowed whole.

When the chicks finally reach the freshwater springs they are reunited with their parents. In a crèche of thousands, finding one another would seem an impossible task but each chick has a unique call and somehow the parents are able to hear this amongst the noise of the others.

VOLCANOES

STUCK IN THE MUD

If entering Fernandina's crater is like taking one's life into one's hands, it's not a great deal different for Lake Natron. Only a few intrepid people have ever tried to cross the central flats of Natron. Leslie Brown, the famous East African ornithologist, almost died in an abortive attempt to reach the flamingo colony on foot. When he managed to make it back to shore, he was at a point of near-death from his struggles with the cloying soda mud and covered with second- and third-degree burns from its chemical toxicity.

Even those crossing the lake by air can be lured to their deaths. The mummified carcasses of migrant birds and bats litter the shores of the lake.

Below **Cameraman Matt Aeberhard** by his filming hide during the flamingo shoot on Lake Natron, Tanzania. To the left is the team's hovercraft, which is the only safe way to cross the toxic lake.

And it's not just animals. In 2007, a helicopter crashed into Natron, killing everyone on board. No one is quite sure why these events have happened, but it's thought the mirror-like surface of the lake creates an artificial horizon that sometimes confuses those who fly over it.

As on Fernandina, filming in Lake Natron takes a great deal of thought and planning. Matt Aeberhard, our cameraman, has filmed in extreme locations all over the world and is one of the few who has actually filmed out on the soda lake. In his opinion, filming on Lake Natron is one of the greatest challenges for any cameraman.

The biggest issue with Natron is access. As Leslie Brown discovered, walking out across the lake on foot is almost suicidal. The only safe way to cross the soda is by hovercraft so purchasing one was our first task. This had to be bought in the UK and then shipped out – a journey that took six weeks. It was an expensive investment – particularly since we didn't even know whether the flamingos would nest on the lake within our production period.

Once the hovercraft was in Tanzania, assistant producer Darren Williams monitored satellite shots of the water levels on a weekly basis and discussed the lake conditions with a flying doctor service that regularly flew over the area. This went on for a year. Then, when it was almost too late for the programme, the water level hit the right mark and the flamingos flooded in. Within a couple of weeks, Matt and Darren were out on Natron filming what proved to be one of the largest nesting events in years.

Below **Matt Aeberhard** rearranges a collection of cardboard cut-out flamingos around his filming hide. Out on the salt flats, the fake flamingos help to camouflage the hide, which make the birds much less wary.

It wasn't all plain sailing. The hovercraft required constant maintenance: its fabric skirts, which give it lift, were torn apart by the sharp soda. Even the most innocuous salt flats were like a belt sander on the integrity of these skirts. Consequently, they needed to be repaired every few days. Fortunately, a group of local Maasai ladies came to our rescue with needle and thread and kept the skirts in working order throughout the shoot.

Even when the hovercraft was working well there were other problems to overcome, like the constantly changing nature of the environment. No journey out to the colony was the same. Plates of soda can rise out of the water like a man's hand. One day a buckled line of these baked hard soda plates might be just about passable; the next day they would be an impossible barrier. The lake's changeability was not good for film continuity either – one day the wet centre might flush with red algae, the next it could be brown and bone dry.

And the hovercraft was just the means of transport to and from the colony. To film the flamingos, Matt still had to be on foot on the soda, hiding away in filming blinds, whilst slowly sinking into the oily, noxious mud. But as he said, 'If you can deal with the acrid smell of the soda and the sulphurous stink of halophile bacteria; if you can deal with the stinging eyes, the cuts and lacerations, the chemical attacks on the legs from the intense alkali; if you can deal with the snow blindness from the whites of the salt and the mercurial intensity of the sun… if you can cope with all that then you can be rewarded with one of the most spectacular events in nature.'

Below **Assistant producer Darren Williams covered in salt crystals after a hovercraft trip across Lake Natron.**

HELL'S WINDOW

When people think about active volcanoes most imagine a bubbling cauldron of lava at the bottom of a crater. But persistent lava lakes only occur in a handful of places around the world – currently just seven, dotted around the world from Ethiopia to Hawaii, Antarctica to Nicaragua. Last year there were eight, but the one on the island of Ambrym in Vanuatu disappeared after a powerful earthquake. This was the one Rolf and I intended to film in after the Valley of the Geysers.

Pools of lava often appear after an eruption, but these typically dry up and turn to rock within a few days – or, at most, weeks. For a lava lake to be maintained, the molten rock needs to be kept at a high enough heat from volcanic gases like carbon dioxide, sulphur dioxide and steam, from ground water – and this doesn't happen very often.

The largest lava lake is in Mount Nyiragongo in the Democratic Republic of Congo. Currently, it has a diameter of around 700 metres, but the lake's size and depth fluctuate over time. Like all lava lakes, it can be here one day and gone the next – and that's exactly what happened on 10 January 1977. It wasn't until the early 1980s that Nyiragongo's lake reformed.

Mount Yasur on the island of Tanna in Vanuatu is the easiest to see – leaving aside the time and cost of actually getting to Tanna. Yasur is thought to have been active for over 800 years, making it one of the most continuously

active volcanoes in the world. Captain Cook was apparently drawn to the island by the glowing volcano in 1774.

Once at Yasur, you can walk up to the crater rim and look down the steep sides to the lake of boiling lava. It's particularly impressive at night, when the lava glows against the dark background. Every so often the lake erupts, sending lava bombs high into the air. Most of them drop back into the crater, although some do land on the rim and slopes. If any were to hit you it would probably be curtains but, as the volcanologist we were working with said, 'Just keep an eye on the trajectory of any that come close and then move out of its way as it falls to the ground.' This is another reason to view the volcano at night, when the glowing lava bombs are significantly more noticeable.

Standing on the rim, right above the lava lake, you can literally feel the raw power of the volcano. When the lake erupts, shockwaves hit you in the chest. If you watch carefully, you can even *see* these waves. What can't be seen rising up from the crater are the choking volcanic gases, which come and go depending on the wind movement.

Above **Rolf Steinmann filming Yasur's lava lake. With clouds of steam and gas rising from vents on the crater floor, filming windows of opportunity can be short.**

Opposite **A slow shutter speed shot of lava bombs being thrown up by Yasur's lava lake.**

VOLCANOES

PREDICTING AN ERUPTION

So do we know when a volcano will erupt? Yes and no. Some volcanologists compare it to predicting the weather – it can be done but there are caveats. The more history volcanologists have of a volcano the more likely they will be able to predict the possibility of an eruption. And that's one of the caveats – the possibility. It's still extremely difficult to know exactly when a volcano will go off.

The eruption of New Zealand's most active volcano, White Island (or Whakaari as it's also known), on 9 December 2019, which resulted in the deaths of 18 tourists and the injury of many others, is a case in point. A month before, the alert level was raised from one to two, indicating 'moderate to heightened volcanic unrest', but, tragically for that day's visitors, nobody was able to predict the actual eruption. As a volcanologist from the University of Auckland said, 'Even though there was increased activity, there was no sense of what was going to happen.'

Much of what we know about eruptions comes from volcanoes like Vesuvius, in Italy, which is the most continuously monitored volcano in the world – not surprising perhaps given its proximity to the city of Naples, with a population of nearly a million. (Vesuvius also provided the first-ever eyewitness account of an eruption – documented by Pliny the Younger in AD 79.)

The most important clue to an imminent eruption is the state of the magma – though not always. With the White Island eruption, it's thought no magma was involved. Nevertheless, before a volcano erupts, in most cases magma rises to the surface, breaking rock on the way. This causes minor earthquakes. So any change in quake activity could indicate a likely eruption. Rising magma also causes the ground above to swell – changes that can be detected with GPS. On summits that are either inaccessible or too dangerous, monitoring can even take place from space using satellites to measure infrared radiation and changes in heat activity – both signs of a possible eruption. And, recently, evidence has shown that measuring the levels of CO_2 can help predict a sudden increase of volcanic activity.

In 2019, a revolutionary technique was developed that could completely change how we predict volcanic eruptions. With the data from volcanoes, supercomputers are now being used to run hundreds of models simultaneously, markedly increasing the accuracy of eruption forecasting. Long term, it is hoped that this technology could monitor each volcano every day, updating the state of play like a weather forecast.

Despite these technological advances, the biggest danger still comes from volcanoes that are not being currently monitored or are just totally off the radar, such as the Chaitén volcano in Chile, which erupted in 2008 after more than 9,000 years of remission.

Opposite **A large eruption on Mount Etna, which dominates the Mediterranean island of Sicily and erupts frequently. It is Europe's largest and most active volcano.**

BOOM AND BUST

We can't live without volcanoes but there have been several times in Earth's history when life has struggled to survive because of them. The planet has experienced five mass extinctions, and volcanoes have played a part in almost all of them. Indeed, they were front and centre of the greatest extinction of all, the end-Permian extinction, when, over a period of 60,000 years – a mere blink of a geological eye – the Earth lost nearly 90 per cent of both its terrestrial and marine species. The culprit was a volcanic system known as the Siberian Traps in what is now central Russia, which caused such sustained eruptions that the lava covered an area the size of the USA, 500 metres deep. More significantly, it resulted in the release of massive amounts of greenhouse and toxic gases, which poisoned both land and oceans and caused an estimated 10-degree rise in the Earth's temperature.

And it wasn't just the Permian. There's strong evidence to suggest that sustained volcanic activity – in this case India's Deccan Traps – was the nail in the coffin for the dinosaurs during the Cretaceous period. Everyone knows that the asteroid that hit the Yucatán Peninsula sent the dinosaurs into a tailspin, but they may well have pulled through were it not for the Deccan Traps. The theory goes that the asteroid set off a massive increase in volcanic activity, lasting 300,000 years, which had the same effect on the climate as the

Below African savannah elephants in Kenya's Amboseli National Park, with Tanzania's dormant volcano Mount Kilimanjaro – Africa's highest mountain – in the background.

Permian's Siberian Traps. So, if the asteroid was the gun, then the Deccan Traps were the bullets. The only silver lining to this event was that the demise of the non-avian dinosaurs led to the rise of the mammals... and us.

It is no coincidence that the development of our sophisticated human culture has taken place in a period of volcanic and climatic stability – an era known as the Holocene. For the last 10,000 years, life has existed in harmony with volcanoes (leaving aside the short-term chaos caused by eruptions like Tambora and Pinatubo) and it's likely this state of affairs would have continued for a long time to come – were it not for the carbon we are now pumping into the atmosphere. The mass extinctions of the Permian and Cretaceous show the dangers of carbon dioxide, and yet we are putting 60 times more CO_2 into the atmosphere than all the volcanoes on Earth today. You could say humans are the new volcanoes – the consequences of which could lead to a sixth mass extinction.

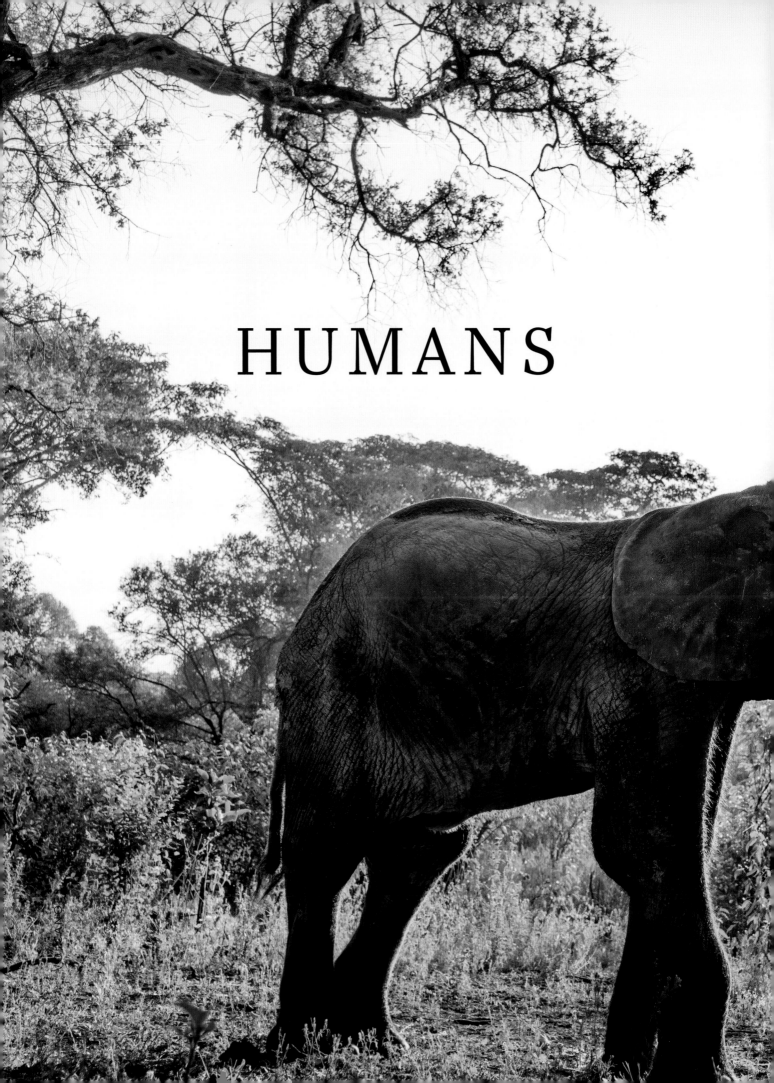

HUMANS

A NOT SO PERFECT PLANET

The forces of nature have been busy of late. On 14 March 2019, Cyclone Idai hit Mozambique, killing hundreds and causing around $2 billion of damage, including the complete destruction of 780,000 hectares of crops in Malawi, Mozambique and Zimbabwe. It was the third deadliest cyclone on record. Five months later, on 28 August, tropical storm Dorian slammed into the Caribbean islands. Wind speeds of nearly 300 kilometres per hour made it a category 5 mega-hurricane and the second strongest Atlantic hurricane on record, bringing an estimated $9 billion of damage. A couple of months after that, on 12 October, Typhoon Hagibis made landfall in Japan, triggering catastrophic flooding, killing 98 people and inflicting damage estimated at more than $15 billion, making it the second most expensive typhoon in history – and continuing a recent trend. Since 1950, three of the most destructive Japanese typhoons have occurred in the space of two years.

Powerful winds also made themselves felt in America's Midwest, which had its most active tornado season for a decade in the same period. In May 2019 alone, there were 556 confirmed tornadoes – the second highest number on record for a single month.

In November 2019, Venice experienced its worst flooding in over 50 years – the second highest water level on record – which inundated over 80 per cent of the city. In the same year, major flooding also affected many other countries around the world, including Argentina, Australia, Canada, the United States, Uruguay, Iran and the United Kingdom, which suffered some of its worst flooding in recent years.

Below **In 2019, Cyclone Idai raced towards Mozambique and Zimbabwe. It was one of the deadliest tropical storms on record to have hit Africa.**

Above **An aerial view of the aftermath of Cyclone Idai, and the devastation it caused near Bebedo, Mozambique.**

It's been hot, too. In Europe, the 2019 June heatwave was the greatest June heatwave in the continent's history, resulting in France and Andorra recording their highest temperatures since records began. Paris had 34 consecutive days without rain – the longest dry spell on record. The blistering temperatures of June were quickly followed by the July heatwave, which was the most intense in European history. Belgium, Germany, Luxembourg, the Netherlands, Norway and the UK all recorded their hottest temperatures ever. Countries in the Southern Hemisphere didn't fare much better.

New Zealand, for example, experienced record high temperatures – as did Australia, where the average summertime temperature was the highest on record by almost 1°C. Australia also had its hottest day in recorded history (in fact, nine of their ten hottest days since records began occurred in 2019), and the highest temperature reliably measured on Earth in any December – a staggering 49.9°C, recorded in Nullarbor in Southern Australia. The optimum air temperature for the human body is between 18°C and 24°C; beyond that, there's an increasing danger of dehydration and death. About 50°C is considered the absolute limit for human survival as, at these temperatures, the body finds it impossible to cool down quickly enough to counteract the effect of the heat.

In California, 7,860 fires were recorded, resulting in over 100,000 hectares of burned land. Hot, dry weather also led to some of the most apocalyptic fire activity ever seen in Australia, which burned 15 million acres and caused huge collateral damage to its wildlife. It is estimated that anywhere between half a billion and one billion animals died in these fires.

Above After recent bushfires, a koala in a eucalyptus tree has a close-up health check from vets on an aerial work platform at Gelantipy in the East Gippsland region of Australia.

Opposite A mother eastern grey kangaroo and her joey are surrounded by burnt forest. They were the survivors of a huge bushfire at Mallacoota, Victoria, Australia.

So 2019 was clearly a tough year for humans and nature, but could you say these events were just coincidence or even bad luck? After all, hurricanes, floods and fires have always happened. Indeed, these events are entirely natural. Hurricanes, for instance, help to transfer heat from the equator to the poles, and Australia's eucalyptus forests are essentially designed to burn. What isn't natural, however, is the frequency and severity of these events. Extreme heatwaves and floods, which used to be once-in-a-century occurrences, are now happening somewhere around the globe every year, or every couple of years. You only have to listen out for the phrase 'since records began' – now being used with worrying regularity – to realise how often we are experiencing severe events. The statistics are certainly not encouraging.

The six warmest years across the globe since 1880 were the most recent six. Using the 1980s as a starting point, each successive decade has been warmer than the last. In 2019, the average annual temperature was 1.1°C above the pre-industrial conditions of the nineteenth century. This may not sound much, but you have to go back a long time in history to find a time that was hotter – 10,000 years, in fact, according to data from polar ice cores.

Above **A bushfire rages in a spotted gum forest on the south coast of New South Wales, Australia.**

You can see what impact these rising temperatures have had on, for example, Arctic sea ice. Since records began, the past 13 years have seen the least amount of sea ice. Scientists now predict that the odds of a totally ice-free Arctic Ocean during the summer will be around 50 per cent by 2030. And, with no ice, there's nothing to reflect the summer sun back into space, which will make the planet even warmer – equivalent to about 25 years of emissions from burning fossil fuels at current rates. These feedback loops, also seen in

Overleaf **Crab-eater seals crowd onto broken sea ice off the Antarctic Peninsula in mid-summer.**

the exponential melting of the Arctic permafrost, which is estimated to store as much as 1,600 gigatonnes of carbon, or nearly twice what's currently in the atmosphere, will hasten the fabled tipping points, beyond which there may be no way back for our planet's climate – at least in a human context.

The seriousness of our current situation is why this is now the most important story of our time.

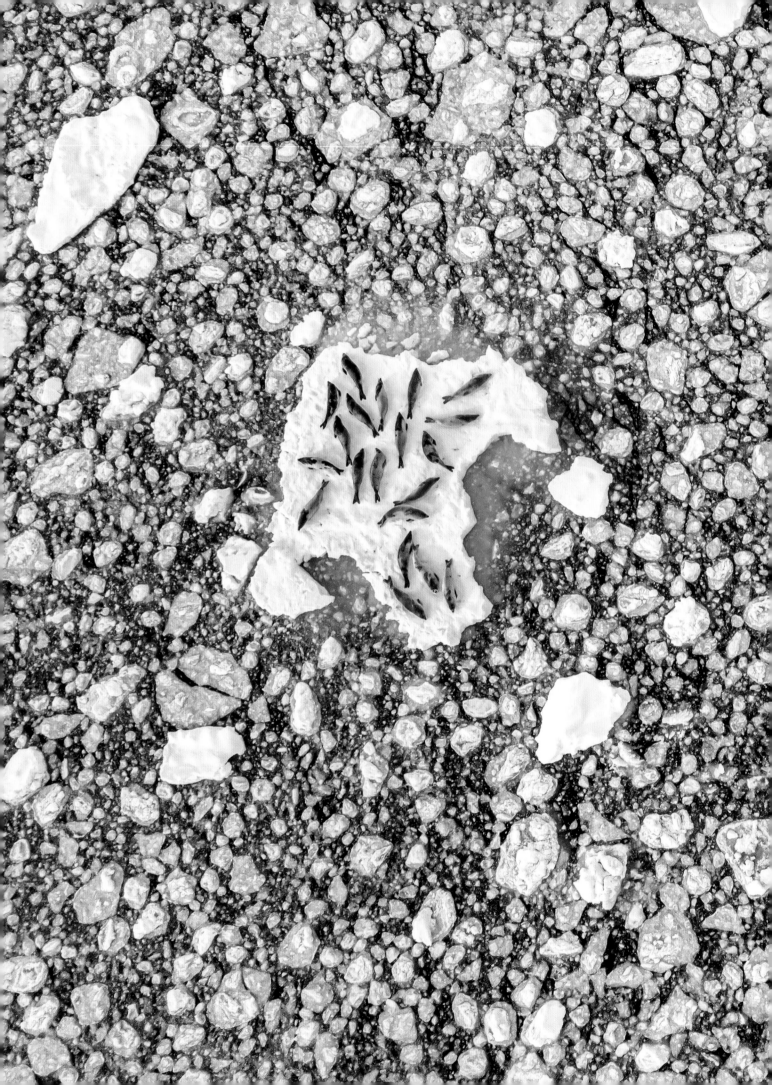

THE INVISIBLE GAS

The evidence of a rapidly warming planet is clear, and the responsibility for it is undoubtedly down to us. Our impact on the planet is now so great that scientists have suggested we have entered a new geological era: say goodbye to the Holocene, during which the planet recovered from the last of the ice ages, and welcome to the proposed Anthropocene, when humans have become the major force on the planet, and not necessarily for its good. In fact, we are now one of the most numerous mammal species on Earth (second only to the brown rat) – a rate of increase that has been particularly dramatic over the last century.

In 1803, after around 200,000 years of modern human history, the world's population eventually reached one billion. It took just another 124 years to reach two billion. By 1975, the population had hit four billion, and 12 years later five billion. (That's worth saying again – 200,000 years to get to one billion, but just 12 years to add an additional billion and go from 4 to 5 billion.) Fast forward another 13 years and the total population of *Homo sapiens* has increased to 7.6 billion. Although our population growth rate is now on a gradual trajectory downwards, we are currently increasing at around 1 per cent a year. It means that the global population is estimated to reach 11 billion by 2083.

Inevitably, this rapid increase of people has come at the expense of wildlife populations. Since 1970, humanity has wiped out 60 per cent of all mammals, birds, fish and reptiles. Today, 70 per cent of all the birds on the planet are poultry and 96 per cent of the mammals on Earth are livestock and humans. Only 4 per cent are wild mammals.

Below **Cooling towers dominate the landscape at the Ratcliffe-on-Soar coal-fired power station, Nottinghamshire, UK.**

The steady loss of wild animals is, perhaps, hardly surprising given we have cleared forests, turned grassland over to agriculture and livestock, drained swamps, overfished the oceans and more; but, there's a much bigger threat on the horizon, and it comes from the way we create energy.

Almost every facet of modern human existence relies on the burning of fossil fuels – whether for transport, heating or in the manufacture of the items that fill our houses. As the American economist and conservationist, Jeremy Rifkin says, 'Here's the problem. Over the last two centuries we dug up the burial grounds of a previous geological era in history – the Carboniferous era – and we took those dead remains in the form of oil, gas and coal. We made an entire industrial civilisation based on these fossil fuels'; and, we now know this has come at a big cost.

When volcanoes erupt, when fires blaze, or when we burn fossil fuels, an invisible, odourless gas is released into the atmosphere. It's called carbon dioxide or CO_2: one carbon atom bonded to two oxygen atoms. Every plant on Earth depends on this gas to grow, absorbing it from the air during photosynthesis. This carbon cycle is, in fact, fundamental to life on Earth – providing the building blocks for all living things. Since the Earth is a closed system, the amount of carbon on our planet never changes, but where it's found does vary. Most carbon is stored in rocks and sediments, the rest is in our oceans, living organisms or – and this is the important point – in the atmosphere, as carbon dioxide. When there's more carbon dioxide than life can absorb, it builds up in the Earth's atmosphere – and it's become increasingly clear that we mess with the planet's carbon cycle at our peril.

Globally, we emit over 36 billion tonnes of carbon dioxide per year into the atmosphere – a figure that has been increasing year on year. In 2018,

concentrations of CO_2 reached new highs – 412 parts per million (an increase of 125 parts per million since the Industrial Revolution). This figure is the highest level it's been for several million years. The last time it was this high, sea levels were 15-20 metres higher. The reason these concentrations are so dangerous is down to the properties of carbon dioxide. CO_2 is a so-called greenhouse gas – that is to say it absorbs heat from the sun, and with less of this solar radiation bouncing back into space, the result is that the planet gets warmer. And this warming planet is threatening the delicate balance of our natural forces, like the weather or ocean currents.

Indeed, the evidence is now so overwhelming that only the most obstinate, or deluded, would deny that human activity is behind the change in our climate. And if we want to glimpse the future we only need look back in time.

In our planet's 4.5-billion-year history, it's been through at least five mass extinction events, and soaring carbon dioxide levels from cataclysmic volcanic eruptions were a significant factor in most of them. The greatest extinction event on the planet so far occurred during the Permian era, 250 million years ago. It was caused by the superheating of the world through massive volcanic activity in an area known as the Siberian Traps, which pumped huge amounts of carbon into the atmosphere – massively raising the global temperature. It led to the destruction of as much as 90 per cent of life on Earth.

Today, humans release 100 times more carbon dioxide than all the volcanoes combined – and at a rate that's even faster than the Siberian Traps of the Permian era. It seems that we are now acting like a supervolcano. As a consequence of this – and the other damage we are doing to the planet through habitat destruction, over-exploitation of wildlife and pollution – scientists have estimated that we are losing species at 1,000 to 10,000 times the normal rate, with multiple extinctions daily. It's why many experts believe we are now going through the sixth mass extinction.

Opposite **An industrial plant at one of the world's largest producers of synthetic crude oil from oil sands, located in Alberta, Canada.**

Below **Youth Climate March, Houses of Parliament, London, 2019. During this demonstration, thousands of children lobbied the government to take urgent action to reduce global warming.**

WHEN DROUGHT BITES

With global weather patterns becoming increasingly erratic, it's ramping up the challenges for wildlife. In Africa, in particular, climate change is leading to less frequent and more unpredictable rainfall. As we rapidly warm the Earth, we're creating severe droughts, threatening the survival of many animals – and no more so than one of the planet's most iconic species, the African elephant, already under huge pressure from decades of habitat loss and poaching.

In Kenya, a prolonged drought over the past few years, the worst since the 1970s, has led to the deaths of hundreds of elephants. When the rains fail and waterholes and rivers dry up, so too does the food and water. An adult elephant needs around 200 litres of water a day, as well as up to 200 kilograms of food, so the victims are often elderly elephants or females with young that find themselves tied to drying water sources. You can see the scale of the problem in the dozens of orphaned baby elephants that are rescued by the Sheldrick Wildlife Trust.

'When we find them', said Angela Sheldrick, 'they are in a pretty sorry state – physically and psychologically damaged after losing their mothers and the protection and support of the herd.' Some are literally at death's door when they arrive at the orphanage. During our shoot, a two-week-old baby elephant came in that required medical intervention, including days on a drip. Nobody knows exactly what had happened to its family, but the nearly newborn elephant was found on its own suffering from burnt feet and sunburn across its body. Fortunately, it has since made a full recovery and is now in the long process of learning the skills it will need to survive back in the wild.

At the orphanage, the baby elephants are looked after by a dedicated team of keepers, who provide round-the-clock support. According to Nick Shoolingin-Jordan, the *Humans* producer and director, the huge effort required to care for a baby elephant is not much different from looking after a newborn human baby (he had just become a father again so he was in a good position to judge).

Eight times a day – sometimes more – the keepers make up a large, 2–3-litre bottle of milk for each orphan. In between, they help the baby elephants burn off excess energy through play. The keepers even sleep in the same room as their charges, covering the youngsters with a blanket to keep them warm during the cold nights. This 24/7 care continues until they are four or five years old, after which they are ready to return to the lands of their birth and an independent life. So far, the Trust has released more than 150 orphaned elephants back into the wild but, to survive, these now need to live in managed reserves, where people must top up water supplies when drought bites again.

Sadly, for Angela Sheldrick and her team, saving elephant victims of drought is not likely to end any time soon. As Angela says of the situation in Kenya, 'Over the years we've seen an enormous change in the weather patterns – much greater unpredictability with dry seasons getting drier and longer.' She echoes the feelings of many who have had a long and close relationship with nature. 'It's the eleventh hour now. We have just one home and we, as the dominant species, should take care of it – must take care of it. It is our responsibility.'

Opposite **Two orphaned elephants with their keepers from the Sheldrick Wildlife Trust, Tsavo National Park, Kenya.**

THE GREAT GREEN WALL

Of our natural forces, the weather is the one that is most noticeably changing before our eyes. The reason is simple: a warmer planet holds more water in its atmosphere. In fact, for every degree that the temperature goes up because of CO_2 emissions, the atmosphere sucks up 7 per cent more water. The result is more concentrated rainclouds and more extreme, unpredictable and out-of-control weather events, such as extreme category 5 hurricanes and severe flooding. As illustrated at the start of this chapter, it's not just wind and rain but heat, drought and fire too.

This is not just affecting wildlife but us as well. Changing weather patterns mean, for instance, that an ever increasing amount of land is turning to desert, with disastrous consequences for those living in these marginal regions. Niger, in central Africa, is thought to be losing 100,000 hectares of arable land a year to desertification.

It seems the least polluting countries are the most impacted, but the least equipped to deal with the outcomes of climate change. The Sahel region in western and north central Africa is a case in point. It's on the frontline of our warming planet, with temperatures rising around one and a half times the global average. As a result, it's experiencing persistent drought, a lack of food

Above **The Great Green Wall** alongside the western fringes of the arid Sahel in Senegal, West Africa.

and rising conflict over dwindling natural resources. And the situation could worsen still since it's estimated that temperatures in the Sahel region could climb by up to 6°C by the end of the century.

The Sahel is also experiencing unprecedented levels of violence and insurgency, and, while this isn't necessarily a direct consequence of climate change, a growing number of examples from around the world are showing a strong correlation between the two. This general instability is triggering one of the greatest migrations in history. According to Jeremy Rifkin, we are going to see 'millions, tens of millions, and unfortunately hundreds of millions of people migrating from areas that are no longer liveable in the next 50 years'.

In Senegal, producer Nick Jordan filmed in Goulokum Tegg, a particularly hard-hit village. It was a place he referred to as the 'no men village', because almost all the men of working age had left to find employment in the nearest cities – or even much further afield in other countries. The survival of the women and children remaining behind is now dependent on the money these men earn and send back, though even this isn't enough. The village must rely on charitable donations and food handouts from NGOs to see them through the year. Most days the children have to make do with one small meal.

The land the village owns is simply not fertile enough to grow sufficient crops to feed the community, but it clearly hasn't always been like this. Chief Seck, the village leader, is in his sixties and has lived in Goulokum all his life. He told Nick that they were able to grow a huge variety of crops here when he was a young man. Not any more. Walking out across the village's ancestral fields, the chief stopped, scooped up a handful of the sandy soil and watched as it flowed through his fingers, like sand through an egg timer.

Below **Women and children from the village of Koyli Alpha, near the Great Green Wall, in Senegal.**

Chief Seck knows there is no future for his village. Before Nick left, the chief made an emotional appeal, on camera, to world leaders. He said that his people had farmed these lands for hundreds and hundreds of years – and that is all they wanted to carry on doing – but they were now innocent victims of changes they had nothing to do with. He had a point: the average British person will have produced more carbon dioxide in the first two weeks of this year than a citizen of any one of seven African nations does in an entire year. Indeed, this region emits less than 3 per cent of the USA's total greenhouse gas emissions.

Goulokum may not survive as a viable community for much longer, but a remarkable project is showing that there is a way to halt the worrying trend towards desertification, and the march south of the Sahara desert. Running for 8,000 kilometres across the semi-arid Sahel, from Senegal in the far west to Djibouti on the east coast, the project's aim is to plant over a billion drought-resistant trees, such as acacias and baobabs. Senegal has already planted 12 million and the hope is that this Great Green Wall – as it's now known – will save around a million square kilometres of currently degraded land.

The trees stop topsoil being blown away, and their roots trap whatever rain falls. In Senegal, the benefits are already evident. Water wells are filling again and struggling communities are beginning to thrive once more as food security improves, which in turn brings jobs and stability to people's lives. And the trees do something else which we all benefit from: as they grow they remove carbon dioxide from the air. By 2030, the trees of the Great Green Wall will have locked up around 250 million tonnes of carbon. At the moment it is 15 per cent complete but when finished it will be one of the largest living structures on the planet – three times the size of the Great Barrier Reef.

Opposite **Chief Seck from the village of Goulokum Tegg, Senegal, with producer / director Nick Shoolingin-Jordan. Goulokum has been particularly hard hit by the effects of climate change. Today, their ancestral lands are no longer sufficiently fertile to grow enough crops to feed the community.**

Below **Tending the vegetable gardens in the village of Koyli Alpha, near the Great Green Wall, Senegal.**

DIVERSITY CENTRAL

Trees have always been an important store of carbon. At present, the world's forests hold as much as 45 per cent of all land-based carbon. For several million years, one of the greatest carbon sinks on the planet has been the Amazon rainforest – an area of 5.5 million square kilometres and, until recently, blanketed with around 400 billion trees. It's a natural treasure producing around one-fifth of the Earth's oxygen and deflecting a significant amount of heat from the sun. At the same time, it is thought that as much as 20 per cent of the world's fresh water cycles through its rivers, plants, soil and air. In fact, the forest is so big it generates its own rain through a process known as transpiration, and this water carries as far as Argentina and the American Midwest.

The Amazon is also the most species-rich forest on the planet, containing as much as 10 per cent of the Earth's known species, including around 40,000 plant species (three-quarters of which are unique to the Amazon), more than 400 kinds of mammal, 1,300 different types of bird and, with new species being discovered all the time, probably well over two million species of insect. These animals are vital to the health of the forest as it depends on them for pollination and seed dispersal, as well as keeping leaf predators in check. In fact, the incredible diversity of the Amazon helps to ensure that no one species gets the upper hand, and a diverse forest has other advantages too.

Opposite **A rainbow appears over cloud forest at Cosanga, in the Napo Province of Ecuador.**

Below **An Amazonian horned frog is camouflaged amongst leaf litter on the floor of lowland tropical rainforest in the Manu Biosphere Reserve, Peru.**

Research has shown that a jungle rich in animal species stores more carbon than one without. And the animals themselves are not an insignificant store of carbon, which is used to build muscles, bone, fur, feathers and keratin. When each animal dies, part of this store of carbon is transferred to other animals, when it is eaten, or into the soils.

Sadly, this great forest is increasingly under threat. About 20 per cent of the Amazon has already been lost through human activity – and it's yet another destructive trend that seems set to continue. The scale of the current deforestation is shocking. Imagine the size of a football pitch – an area of about 100 metres by 70 metres – because that is how much of the Amazon is being lost through deforestation every 30 seconds. That's the equivalent of 2,880 football pitches a day! Today, around 11 per cent of global greenhouse gas emissions can be blamed on deforestation directly caused by humans – that's roughly equal to the emissions from all the cars and trucks on Earth. Or to put it another way, if tropical deforestation were a country, it would rank third in carbon dioxide equivalent emissions behind China and the USA.

In Brazil, which holds around 60 per cent of the Amazon, one-fifth of the forest – or around 780,000 square kilometres – has been cleared for agriculture and other development projects over the last 50 years. The biggest jump in

Above **Satellite view of many forest fires on the southern edge of the Amazon rainforest in Brazil.**

HUMANS

forest loss occurred between 1975 and 1985 but it has continued at around 14,000 square kilometres per year – that is until 2019, when deforestation rates jumped 50 per cent from the same period in 2018 – half of it in supposedly protected areas, including hundreds of indigenous tribal lands that cover around a quarter of the Brazilian Amazon. With the national government rolling back laws on land use restrictions and actively encouraging the exploitation of the forest for agribusiness, hundreds of thousands of hectares have been felled or burned in the last couple of years. In 2019, the National Institute of Space Research reported a staggering 80,000 fires from the Amazon – the majority of them lit on purpose by loggers and land grabbers. This was an 84 per cent increase from the same period in the previous year – big enough to attract the world's attention and put pressure on the Brazilian government to halt the destruction.

While filming for this and other episodes of *A Perfect Planet*, crews said smoke from burning forest was ever present – even in the remotest corners of the Amazon. Producer Daniel Rasmussen spent several weeks filming in Brazil for the *Humans* episode and witnessed some of this destruction first hand: 'Everywhere we drove and flew, fires were raging,' he said. 'People stood brazenly on roads with fire starters meticulously erasing areas of the forest.'

Deforestation on this scale is already having an impact on the Amazon's weather – something that indigenous people started to become aware of even decades ago. Across the western Amazon, for example, there has been a noticeable delay to the onset of the rainy season, which has also become

Below **Tropical rainforest is burned to clear the land for cattle ranching at Pará, Brazil.**

shorter and warmer. Scientists worry that if the Amazon was to lose a further 20 per cent of its forest – entirely possible at current rates – then there could be a real possibility of a tipping point being reached whereby the forest can no longer generate enough moisture to maintain the habitat, leading to a transition from forest to savannah.

Most people would agree that protecting the planet's diversity of plants and animals is a good thing in itself. After all, we are the only planet we know of with life, and the astonishing variety we have on Earth should be cherished, with an emphasis on preserving the ecological integrity of every habitat. This ethical value, sadly, hasn't been enough to persuade those with the power over our wild spaces to protect them – and the Amazon is no exception.

Working out the true value of every intact forest, however, could be the key to protecting them. It has been estimated, for instance, that one hectare of livestock or soy is worth between $25 and $250, while the same hectare of sustainably managed forest can yield as much as $850. These economics have yet to win over decision-makers in many Amazonian countries but a new cutting-edge technology just might. It's led by a team from the Global Airborne Observatory. They've developed a way to quantify how much carbon a forest stores. By firing high-powered lasers across the canopy at 500,000 times per second, they can literally map the amount of carbon each tree holds. These maps enable countries to see the real value of their forests, and with the help of the international community – whether through countries or large multinationals – pay governments to keep them standing, for instance, as part of carbon offsetting schemes. Already, this work has shown that we've been underestimating carbon stocks and emissions in the Peruvian Amazon.

Preserving the Amazon's remaining tracts of forest is a vital goal in helping to mitigate the effects of our warming planet – as well as saving indigenous communities across the region and countless species of plants and animals – but so too is replanting cleared or heavily degraded forest.

Above left and right
Milene Alvish and her team scatter the seed mixture over acres of burnt and degraded land.

Growing a forest is one thing, however; *re*-growing a species-rich jungle, like that found in the Amazon, is quite another and has always proved difficult. So, one reforestation project has turned to the indigenous peoples of the Amazon who have helped in a revolutionary approach to this problem. Its aim is to plant a new jungle of 70 million trees, covering 30,000 hectares, making this project the largest tropical forest restoration in the world.

The Xingu Seed Network hires indigenous people and local farmers to collect seeds – using their unique knowledge of the forest to source seeds from the Amazon's most important tree species.

By mixing the seeds of at least 200 different kinds of tree together, they create a super recipe known as a muvuca, which is then spread over every square metre of burnt and mismanaged land. To plant one hectare of new forest involves scattering as many as 200,000 seeds from native trees, and this technique is succeeding where others have failed, helping to jump-start a new, healthy and diverse jungle. Climate change may not be solved by planting trees, but it can still make a significant difference.

THE GREAT CARBON STORE

The carbon dioxide produced by humanity is attacking another crucial part of our planet – the oceans – and life cannot survive without these vast bodies of water, which cover over 70 per cent of the Earth's surface. They are thought to produce around 70 per cent of our atmospheric oxygen – thanks to the work of photosynthetic phytoplankton. They also help to regulate the planet's temperature, and it is estimated that they absorb around 90 per cent of the heat our carbon emissions have trapped in the atmosphere. But, perhaps most importantly of all, they are a vital long-term carbon sink, having absorbed around 40 per cent of CO_2 emissions since the start of the industrial era. The oceans, in fact, are far and away the planet's biggest store of carbon, with an

Above **Stony corals bleach on Heron Island, at the southern end of Australia's Great Barrier Reef.**

estimated 40,000 billion metric tonnes of carbon in the ocean itself. As on land, all living things in the oceans, from plankton to whales, store carbon in their bodies. When these creatures die, they drift to the ocean floor and enter the carbon cycle again. The carbon that isn't released into the water through decomposition builds up in the silt, where, over millions of years, it gets compressed into sedimentary rocks. These rocks store even more carbon than found in the ocean – as much as 100 million billion metric tonnes.

At the base of the ocean food chain are phytoplankton, tiny plant-like organisms, which everything else depends on. Without them, the entire food web would collapse. Yet phytoplankton are now under threat thanks to

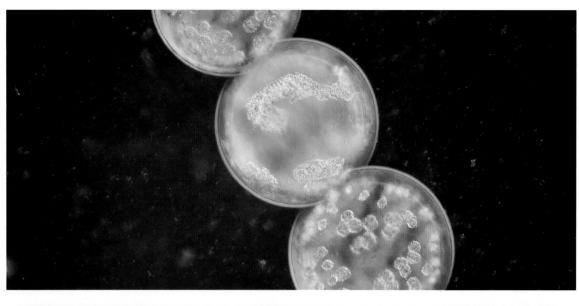

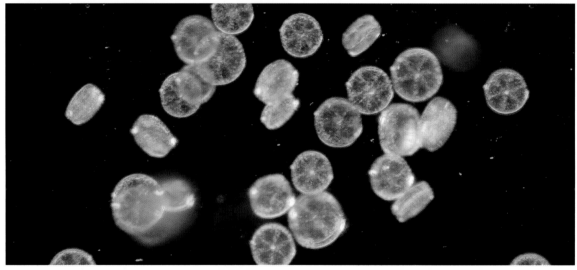

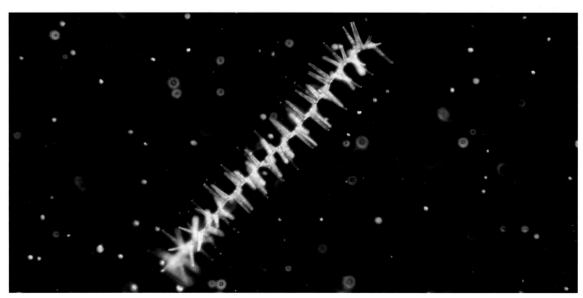

human activity. As the oceans absorb more and more of our emissions they're becoming warmer, which is separating the surface waters from the supply of nutrients below – nutrients the plankton depend on. As a consequence, the global population of phytoplankton has fallen rapidly – according to research estimates, by as much as 40 per cent since 1950.

If the warming of the oceans wasn't bad enough, many kinds of sea life are also under pressure from something else. When carbon dioxide dissolves in water it makes it more acidic and this acid dissolves carbon, threatening everything with a shell built out of calcium carbonate, including many kinds of plankton, clams, starfish and corals. Acidification could even lead to the disappearance of all coral reefs – one of the most valuable and biodiverse habitats on Earth – since a rise in sea temperatures causes coral to expel the symbiotic algae that provide much of their food. It's this that causes the phenomenon of coral bleaching, which, if it happens repeatedly, results in the death of the coral. In one year, high sea temperatures on the Great Barrier Reef killed off around 20 per cent of its shallow-water corals.

The last time the double whammy of warming and acidification hit the oceans in a big way, back in the Permian era 250 million years ago, it contributed to a devastating mass extinction of life – with as much as 96 per cent of all marine species being lost forever.

OCEAN PATROL

And the human force is hitting the oceans on yet another front. Up to three billion people around the world depend on seafood as their primary source of protein yet this too is under threat after decades of overfishing. It's a state of affairs that is succinctly summed up by Sri Lankan marine biologist Asha de Vos: 'We've thought the ocean is this infinite space that is full of infinite resources and this infinite capacity to withstand and tolerate everything that we throw at it. But it's not and it can't.' We're now discovering that, like everything else on this planet of ours, the ocean has it limits.

Today, around four million fishing vessels crisscross our oceans using increasingly efficient techniques for finding and catching fish. A purse seine net, for example, can trap thousands of fish at one time, and, each year, trawlers plough an area of seafloor twice the size of the continental United States. Since 1950, these techniques have removed around 90 per cent of all large ocean predators, including bluefin tuna, halibut, marlin and sharks. As a result, 17 out of the 39 species of shark living in the open ocean, such as the oceanic whitetip shark, are now threatened with extinction.

A large part of the problem is down to how much of the ocean is protected – currently just a little over 3 per cent. Compare that to the 13 per cent on land. Thankfully, this is beginning to change, with an increase in marine protected areas (MPAs). The first stage is to raise the level of protection to 10 per cent of the ocean's total area, but the ambition is to reach 30 per cent, giving marine ecosystems a chance to recover and helping to make oceans more resilient to climate-related disturbances.

Off the coast of Gabon, they've created one of Earth's most ambitious networks of MPAs – a total of 20 marine parks and reserves which will protect 26 per cent of Gabon's territorial waters across 53,000 square kilometres. This area is a critical hotspot for sharks, sea turtles and breeding whales and dolphins. Studies on MPAs show that with effective protection marine life can, quite quickly, bounce back, even beyond the protected area.

The big issue for many poorer nations – like those along the West African coastline, which has some of the most diverse fisheries in the world – is how to enforce the protection in the absence of any effective navies. Many of the so-called distant-water fishing fleets from Asia, Europe and the USA know that they can fish these waters with little chance of getting caught. As a result, illegal fishing is thought to represent 40 per cent of the catch in West African waters, the highest level of any region in the world. In addition, fishing fleets from China, Europe and the USA are heavily subsidised, which makes it almost impossible for local fisheries to compete and this, in turn, has led to a decline of coastal communities across West Africa.

Given the huge challenge of protecting their coastline, the Gabonese government has teamed up with the conservation group Sea Shepherd, which, with its ship the *Bob Barker*, patrols thousands of square kilometres of the newly gazetted marine park. Just its presence is often enough to deter many

foreign fishing boats from these waters, though not always – particularly when it comes to the huge European tuna vessels which, it is alleged, are able to pay whatever is needed to have all the proper licences and documents. Early on, while taking Gabonese officials on patrols of their water, Sea Shepherd discovered that European tuna boats had been catching up to 60,000 tonnes annually but only declaring a maximum of 15,000 tonnes.

Sea Shepherd also carries out inspections on boats fishing outside the reserves because, while they might have a licence to fish, they could be catching more than their quota, as well as bringing up protected species, such as dolphins and sharks, which are indiscriminately caught in fishing nets – the so-called bycatch. The numbers of this bycatch are staggering. Globally, millions of sharks and 300,000 whales and dolphins are accidentally killed each year. Sea Shepherd patrols, though, are working. In the past three years, they have arrested 50 vessels and inspected hundreds more. By assisting the coastguard in arresting one vessel that was poaching sharks, destined for the still thriving Asian shark fin market, Sea Shepherd was able to save the lives of around 250,000 sharks.

Above **A former whaling vessel, the 52-metre *Bob Barker*, now patrols the coast of Central West Africa on a mission to tackle illegal, unregulated fishing.**

HUMANS

ABOARD THE *BOB BARKER*

Every shoot requires a lot of planning, and the one we did for filming on Sea Shepherd's *Bob Barker* was no exception. Nevertheless, despite all the organisation, it's fair to say the trip didn't get off to a flying start. Arriving in Libreville late at night, the crew – made up of field director Emily Franke, cameraman Paul Williams and sound recordist Tamara Stubbs – got just a few hours' sleep before meeting up with Captain Peter Hammarstedt for breakfast. Still bleary-eyed from all the travelling, they listened as Peter told them that the boat was unfortunately going nowhere – at least, for the time being. Just the day before, the President of Gabon had sacked his environment minister over a scandal involving timber smuggling, and, as a consequence, the Gabonese marines, which were needed to assist the patrol, were grounded. Nothing could be done until a new minister was appointed.

To avoid tipping off any fishing fleets to the presence of the patrol boat, the crew had to covertly board a couple of speedboats at first light and race to the *Bob Barker* – a job made slightly more difficult by their 30 cases of equipment. Once on board, there was nothing else to do but wait for the all clear from the government. But bobbing up and down off the coast off Libreville at least gave the crew a chance to acclimatise to life at sea.

After five days of waiting, the ministerial vacancy was filled and the *Bob Barker* was ready to resume its patrols. Industrial fishing takes place 24 hours a day and, since the patrol ship is more likely to catch illegal vessels fishing in the marine protected areas at night, the film crew had to be ready for

Below **Cameraman Paul Williams climbs aboard the *Bob Barker*.**

Opposite A fisherman aboard a European tuna vessel removes shark bycatch from the nets so they can be returned to the sea. Unfortunately, most are so badly injured they don't survive.

action at all times. So each night they slept in their clothes with a grab bag of life jacket, camera kit and batteries at the ready. Of course, knowing that an alarm call could come at any moment meant a good night's sleep was largely out of the question, and it wasn't just the crew that needed to be prepared for action. Every evening, the *Bob Barker* went into dark mode. Deck lights were switched off and all portholes were covered so the ship could approach suspicious fishing boats without being spotted.

To keep mind and body together while on board, there were daily exercise classes on the helicopter deck (a workout made more challenging by both the rocking of the boat and the wonky slant of the deck, which had taken a battering from a whaling ship on one of Sea Shepherd's earlier campaigns). After a vegan supper (one of three vegan meals a day, which Emily said were amazing given the tiny galley kitchen, where the food was whipped up by volunteer chefs), evenings were spent playing bilingual *Bananagrams* with the crew of the *Bob Barker* and Gabonese marines.

Boarding fishing boats during an inspection required scaling rope ladders from RIBs (rigid inflatable boats) held against the side of the vessel. This was no mean feat in choppy waters, while also weighed down with camera and sound equipment. Fortunately, according to Emily, cameraman Paul turned out to be a natural – whizzing up the ladders like a 'sea monkey' – but this rather sweaty work did throw up another issue back on board the *Bob Barker*. Thanks to a faulty pump, strict water restrictions were put in place until they could pick up a spare part – so after days of getting on and off fishing vessels in the midday sun via swaying rope ladders, there was, said Emily, a distinctive 'eau de poisson' around the living quarters, which the whole film crew were sharing.

During the film crew's time on patrol, the crew of the *Bob Barker* inspected boats of two types: the formidable industrial purse seine ships from Europe, which, at over 100 metres long, are like enormous fish-processing factories, and the badly maintained Chinese trawler fleets, where, invariably, the crews had to endure squalid conditions. The techniques and practices of both shocked the film crew, as Emily explained: 'While the trawlers are devastating in terms of the terrible damage they cause to the seafloor – as well as heartbreaking to witness the horrific working conditions of the boat crews – it was the industrial purse seiners that disturbed me most in their sheer capacity to catch and process such phenomenal quantities of fish every day.'

Purse seine nets are over a kilometre long and up to 250 metres deep. They are set in circles around shoals of fish before being tightened at the bottom like a purse and then drawn up to the boat. Not surprisingly, the nets don't just draw up the targeted fish – for the most part, tuna – but also sharks, turtles and a variety of other species of bycatch. 'We saw dozens of sharks land on deck,' said Emily, 'which were then grabbed by their gills and wrestled off the side of the boat, or suspended by their tails from winches and dropped overboard – many too badly damaged to survive.' Emily also saw a huge leatherback turtle tangled and cut up by the nets being hastily dispatched off the back of the boat, out of sight of the cameras. Sadly, this collateral damage is the reality of many of the industrial fishing techniques going on in the open ocean today.

THE POLAR PUMP

Humans aren't just overharvesting life in the sea, we're also disturbing one of the ocean's most vital forces. Powerful global currents, which start in the Arctic and Antarctic, transport cold nutrient-rich water from the depths to the surface, driving almost all life in the oceans; but this force is also under threat.

When ocean water freezes, forming sea ice in polar regions, the salt is left behind, causing the surrounding seawater to become denser. As this heavier, saltier water sinks to the ocean floor, it draws in surface water to replace it, setting in motion a current, known as the 'global conveyor belt', which circulates nutrients, oxygen and heat around the Earth. Almost every drop of seawater on the planet rides this current – taking a thousand years to complete a full circle. But here's the problem: the ice in the Arctic and Antarctic is melting so fast that it's destabilising the system.

In the Arctic alone, 14,000 tonnes of freshwater are emptying into the sea every second. Ice sheets in Greenland and Antarctica are also losing ice at record speeds. In Greenland, this has amounted to a loss of 600 billion tonnes of ice – equivalent to a global sea level rise of around 1.5 millimetres – and a recent report showed that the rate of melting from Antarctica has accelerated

Below **An Eden's whale, with mouth agape, feeds on anchovies in the Gulf of Thailand. Seabirds swoop in to grab fish escaping its maw.**

threefold in five years. Put simply, the polar ice caps have melted faster in the last 20 years than in the last 10,000 years. It's the same story with the world's glaciers. Switzerland's glaciers, for example, lost around 2 per cent of their total volume in 2019. Over the last five years, the loss has exceeded 10 per cent – the highest rate of decline in more than a century of records.

With all this freshwater pouring into the oceans, resulting in a decreasing amount of sinking saltwater, the polar pump – the engine room of the global conveyor belt – is slowing down, meaning less nutrients, heat and oxygen moving around the planet. As Asha de Vos puts it, 'We are dependent on these large circulation patterns that go on in our ocean – this continuous movement of beautiful cold water coming from the depths, which is chock full of nutrients. Chock full of productivity.'

Recent research on sea surface temperatures has shown that the powerful Gulf Stream, for instance, has slowed down by around 15 per cent since the middle of the twentieth century – with human-made climate change being the prime suspect. The work of some scientists suggests that this globally important current may be at its weakest for at least 1,600 years.

A warmer ocean also holds less oxygen – as much 40 per cent less in some tropical waters – and these warm waters support fewer species of fish. In the last 50 years, dead zones (places where a drop in oxygen causes the

water to stagnate) have quadrupled, a situation that has been exacerbated by human pollution flowing off the land into the sea.

In the Gulf of Thailand, agricultural fertiliser washing into the water has resulted in low levels of oxygen and this has had a dramatic effect on the aquatic wildlife, including its resident population of Eden's whales. These whales typically feed by lunging through shoals of bait fish, but there are now so few fish that launching their 15-tonne bodies through the water isn't worth the energy. To survive, the whales have had to adapt. It's resulted in the development of a new hunting technique – one that requires almost no effort.

Rising head first out of the water, the whales simply open their vast mouths and wait. The panicked fish – mostly restricted to the surface because of the low levels of oxygen – leap out of the water and inadvertently land in the whales' mouths. With this ingenious hunting strategy, Eden's whales have found a way to cope with the pressures they now face but, for many ocean species, the speed of these changes is likely to be too fast, and, for those, extinction may well be the ultimate endgame. We are still a long way from the catastrophic conditions of the end of the Permian era, when deoxygenated ocean waters contributed to the extinction of around 96 per cent of all marine creatures, but this is certainly no time to be complacent.

Above **The global movement of ocean currents means that nowhere is immune to the scourge of plastic. Aldabra in the Indian Ocean is one of the most remote islands in the world, but the prevailing currents throw huge amounts of litter onto its shores. Every one of these washed-up flip flops was collected from within 20 metres of the centre of this photo.**

Opposite **Sea ice formation off the east coast of Greenland in winter.**

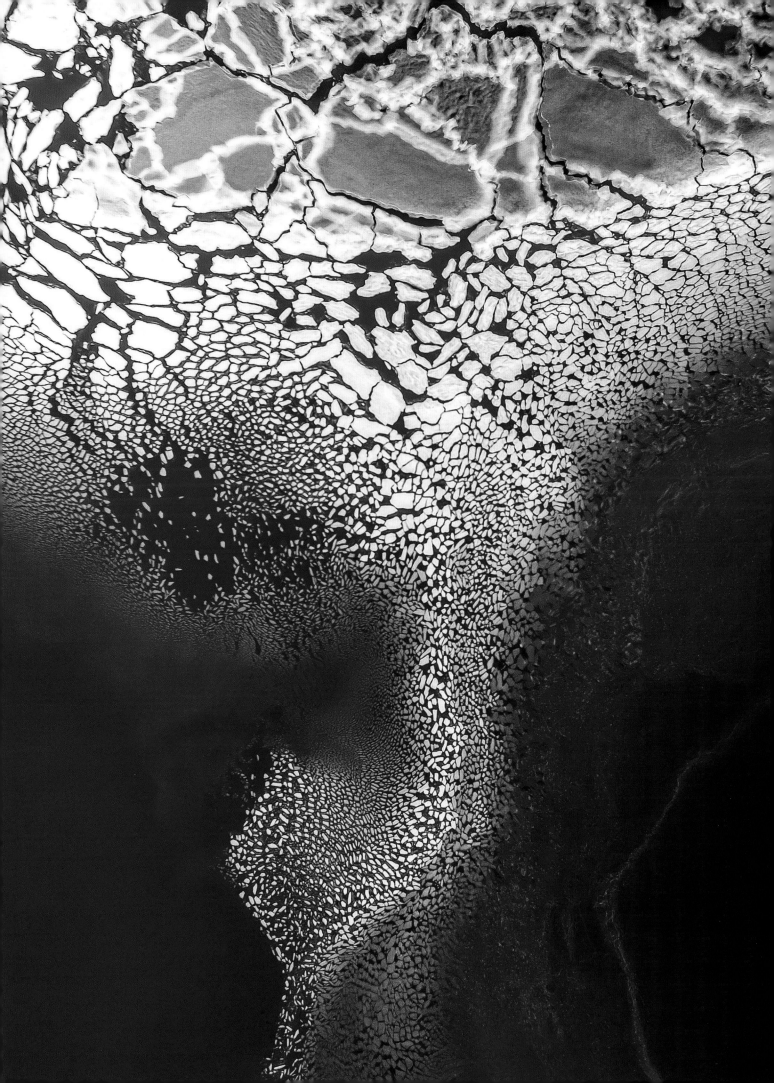

TURTLE TRIAGE

Ocean currents don't just carry nutrients around the globe; they are also highways for many migrating animals. Kemp's ridley sea turtles, for instance, use them to move around the tropical waters of the Gulf of Mexico, and northwards up the coast of America to the Gulf of Maine during the summer. Recently, however, they have been running into trouble – caught out by the changing nature of the currents.

The coastal waters in the Gulf of Maine are warming faster than almost anywhere else on Earth and it has already had an impact on native species like cod, blue crabs and lobsters, which are shifting their ranges northwards, or going deeper, to find cooler waters. The Gulf of Maine's warming waters are also lulling the Kemp's ridley turtles into a false sense of security – causing them to linger much longer in these northerly areas than they would normally. Unfortunately, when the cold autumn temperatures suddenly draw in, the turtles get trapped in water that's too cold for their survival. Disorientated, the cold-stunned turtles either drown or get washed up by the high tide onto beaches. If they're lucky, they will be found by one of the 250 people monitoring beaches at this time of the year as part of the turtle emergency response team, headed up by Bob Prescott, director of the Wellfleet Bay Wildlife Sanctuary on Cape Cod.

In the 1970s and 1980s, fewer than ten turtles would wash up onto the beaches of Cape Cod each autumn. Now, it's common for hundreds to strand – sometimes it's even over a thousand. Most of the casualties are turtles between one and six years old. When they're found, their hearts are often beating at just one to five beats a minute and their blood is barely circulating. They seem as good as dead but, even at this point, there's still a chance of survival. As Bob said, 'It's all about timing. If we can get to them within an hour of them washing onto the beach, then we're going to be able to save as many as 90 per cent of them.' Sadly, however, it's not always so positive. Last year, at Thanksgiving, the number of dead turtles topped 200.

The critically ill animals are rushed to Boston's New England Aquarium where, under Bob's supervision, they are treated in a state-of-the-art emergency room for turtles. Here, the condition of the turtles is assessed – each undergoing a turtle triage. Turtles at death's door are immediately put on ventilators to help them breathe. Almost all are then given stabilising drugs and fluids, after which their lungs are cleared of water and their eyes washed of sand. Producer Nick Shoolingin-Jordan, who filmed the sequence for this episode, said this turtle intensive care unit was like being in a real hospital. 'You have to be really quiet. Everything is done carefully and calmly, to avoid adding any further stress to the patients.'

The vets and nurses need to be extremely careful bringing the hypothermic turtles back to a normal temperature. This starts as soon as they are found. Even the cars the turtles are transported in must not be too warm as the sudden change of temperature can kill them instantly. Once in hospital, however, the

turtles are warmed incrementally from 12°C to 24°C, over around four days. But to bring one of these sick turtles back to full health can take many months of nurturing. It is a lot of effort but certainly worth it for a creature that might live another 50 years – and a species whose numbers have dropped to between 7,000 and 9,000 nesting females worldwide.

When their recovery is complete, the turtles head south and back to the wild. This time by air, on flights that will take them to the warm tropical waters they depend on. Before they leave the aquarium, each turtle is radio-tagged so that scientists can get a picture of their movements and how they are faring in this changing world. To date, only one tagged turtle has returned to the ER unit after re-stranding – so perhaps they learn from their bad experience.

Below A cold-shocked juvenile Kemp's ridley turtle having a blood sample taken, while undergoing life-saving treatment at the New England Aquarium, Massachusetts.

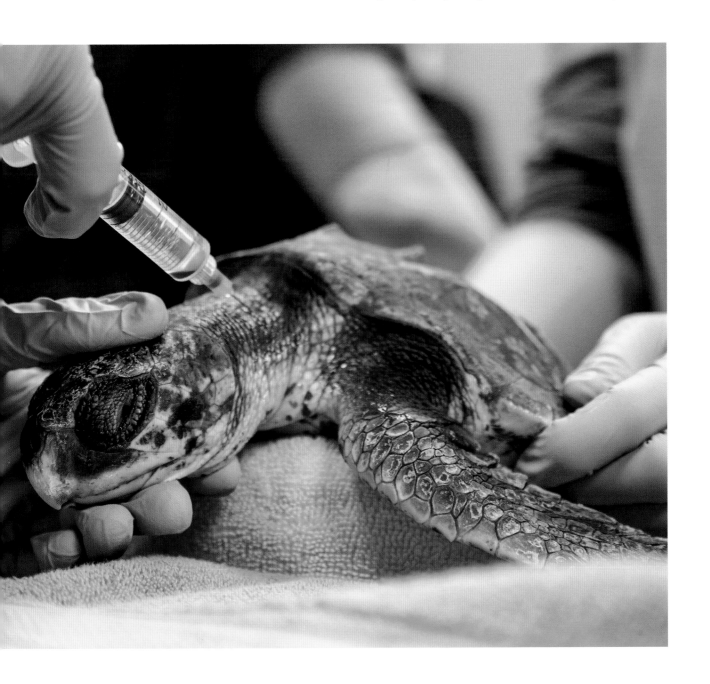

NATURAL FORCES AND THE CIRCULAR ECONOMY

Every serious scientist now agrees that human-produced carbon dioxide is changing the Earth's climate and threatening the delicate balance of our natural forces. Currently, our population of seven billion – and counting – is using the equivalent resources of one and a half Earths. Clearly, this is not sustainable, and the situation is made even worse by the fact that around 80 per cent of the energy we use to power this growing human world comes from burning fossil fuels, which, through carbon dioxide emissions, is altering our weather patterns and destabilising vital ocean systems. Can we reverse the damage we've done and become a force for good? The proposition is simple enough on paper: we need to consume less and generate most of our energy from clean, renewable sources. But this is easier said than done.

In 2015, the countries of the world came together in Paris to agree a reduction in carbon dioxide emissions. The aim of the Paris Agreement was to keep the global temperature rise 'well below 2°C above the pre-industrial levels and pursue efforts to limit the temperature increase even further to 1.5°C'. Unfortunately, these good intentions haven't been enough. The

Below **Morocco's Noor solar power project in Ouarzazate** is the world's largest concentrated solar power complex. Two million mirrors turn sunlight into clean energy that accounts for 6 per cent of the country's electricity supply.

world is seemingly already well on its way to hitting the 1.5°C ceiling – years ahead of schedule. Measurements of carbon dioxide in the atmosphere from key places around the world, such as the 365-metre Amazon Tall Tower Observatory, have backed up this worrying trajectory.

Moving to a carbon-free economy involves a huge change to the *natural order* of things but it's still a no brainer, even on purely financial grounds. As the economist Jeremy Rifkin said, 'The sun has not sent us a bill. The wind has not invoiced us. Coal, gas, uranium. They're expensive. The sun and the wind are free.' Indeed, we could power the whole world with just a fraction of the solar and wind energy we get every year. Yet right now, with only 20 per cent of our energy coming from renewable sources, we have barely scratched the surface, or indeed, below the surface. Currently, we have tapped only 7 per cent of the planet's geothermal potential. Fortunately, things are beginning to change.

Morocco is home to the world's largest concentrated solar farm. Built on an area of more than 3,000 hectares, the Noor-Ouarzazate complex produces enough electricity to power a city the size of Prague. Here, innovative technology is using parabolic mirrors to superheat a fluid-filled pipeline to 400°C. The heat is then stored in a tank of molten salt which allows something that hasn't been possible before: the creation of power during the hours of

Below **The Amazon Tall Tower Observatory in Amazonas State, Brazil. This is one of a number of towers around the world collecting vital data – particularly about levels of carbon dioxide in the atmosphere.**

darkness. Industrial-sized solar power plants like Morocco's are now popping up all over the world, from India, which has four of the top ten largest solar farms, to China, the USA, Mexico and Egypt.

There are also the beginnings of a revolution in small-scale energy production. These are known as *microgrids*, where energy is produced and consumed locally. Already, millions of people around the world are producing their own solar and wind energy in their homes – especially important in rural areas in, say, India or Africa, which might never have been connected to their country's national grid system. Some of the changes might seem simple to us in the West, but they're vital to help bring people out of poverty. One roof-mounted solar panel in a place where there was previously no electricity means, for instance, that children can now do their homework at night. Additionally, the energy not used by the producer can be sold or shared locally, making these communities less vulnerable to power outages resulting from extreme weather events.

As well as creating more energy from green sources, we also need to walk more gently on this unique planet of ours. In other words, we need to recycle more and consume less – particularly those of us who live in the wealthiest parts of the globe, where per capita consumption is at its highest. This will also involve many changes, including shifting our largely meat-based diet to one centred more on plants. (Indeed, veganism is already booming

Above **The Ouarzazate concentrated solar power plant covers an area the size of San Francisco.**

across the western world.) At the moment, 6 per cent of global greenhouse gas emissions come from the raising of beef cattle. Apart from the damaging effect of converting forest to fields, cows – through their digestive process – are big producers of methane, a greenhouse gas that is 25 times more potent at trapping heat than CO_2.

So, instead of the take-make-waste economic model, we need to think in a more circular way – just like nature, where nothing is wasted. In the circular economy, the emphasis is on a low-carbon culture where things are shared and reused, whether it be our clothes, our homes or our cars. To make this easier, new products will need to be designed so they can be easily recycled or upgraded. At the moment, if one of our household items, like a vacuum cleaner or iron, breaks we immediately assume it needs to be thrown away and a new one purchased.

Until we completely separate ourselves from our dependence on fossil fuels, we will all need to make an effort to reduce our individual carbon footprints – after all, using and consuming less means less carbon dioxide is being produced. Doing nothing will have huge ramifications not only for us but for the whole natural world. 'What scientists are telling us,' says Jeremy Rifkin, 'is that we face a runaway cascade of environmental events feeding off each other and taking us into an unknown abyss. This could lead to a very quick mass extinction of life on this Earth in a short period of time.'

Below **Algerian traditional house in the Sahara, equipped with its own solar panel.**

THE FROZEN ZOO

Scientists are not yet in agreement over the current rate of extinction – or whether we have already entered a sixth mass extinction – but all agree that species are going extinct many times faster than they should be, perhaps at the fastest rate since the extinction of the dinosaurs 66 million years ago. In response, and as an emergency backup, institutions in different parts of the world are preserving the DNA of as many animals as possible before they go extinct – a kind of 'Ark of Life'.

San Diego Zoo receives DNA samples from all over the world and preserves them in what they refer to as the Frozen Zoo. Here, in a specialised laboratory, the living cells of the world's rarest animals, and many others, are being stored in liquid nitrogen at minus 196°C. 'It's hard to imagine,' says Marlys Houck, the Frozen Zoo's curator, 'but there is probably more vertebrate life in this room than anywhere else on the planet. We get samples every day. It might be a tiger or a rhino, or a rare reptile. Right now we have 10,000 individuals represented.' The collection includes some animals already on the verge of extinction, like the Northern white rhino or the vaquita porpoise, and the hope is that these samples might provide the opportunity to bring these species back from the dead – and not just these two. Sadly, for far too many other animals, the Frozen Zoo might be needed sooner than we thought.

It might be comforting to know that we can bring species back to life if the worst were to happen, but surely the real solution is protecting these creatures in their natural habitat now. Do we really want to push the natural world to the brink of collapse? Time is certainly running out, but it's not too late. The question is, can we make the necessary changes in time? Perhaps the last word should be left to marine biologist Asha de Vos: 'I see reason to hope. I think we humans are incredibly intelligent animals and we can, and we will do it – if we set our minds to it.'

Opposite **Northern white rhinoceros with its horn removed before being released into the wild at Ol Pejeta Conservancy, in central Kenya.**

Below **Marlys Houck – San Diego Zoo's global curator of the Frozen Zoo – carefully handles a DNA sample of a lemur. Here, the DNA of some of our world's rarest animals is being stored in liquid nitrogen at minus 196°C.**

THE VIRUS AND BEYOND

Humans may be the most dominant force on our planet but the coronavirus pandemic exposed our vulnerability. With most of the world's countries laid low by fears of the virus ripping through their populations and overwhelming health services, economies went into a kind of suspended animation – and, in many cases, a nearly unprecedented reverse. Suddenly, we didn't appear to be the all-conquering and unstoppable species we might have thought, having been halted in our tracks by an invisible enemy that is neither alive nor dead but able to quickly replicate in the living cells of its host.

Some people saw the pandemic as the natural world getting its own back; after all, Covid-19's origins were almost certainly the result of an environmental issue – the misuse of wild animals. Nobody knows for sure how the virus started, or how it made the jump to humans, but it seems to have had its beginnings in the wet market of Wuhan, in China, where, amongst other products, wild animals were traded. Scientists believe that by far the most likely vectors were bats, which were sold as a food item in Wuhan. Whether this proves to be the case or not remains to be seen, but the history of other deadly viruses, such as SARS, HIV or Ebola, shows the dangers of unnatural interactions with wild animals.

The question is whether anything good will come out of this disaster – particularly in terms of the environment and our attitude to climate change. The evidence for optimists is promising. In a survey of the world's biggest economies, on average, 70 per cent of the respondents agreed that, in the long term, climate

Below **A traditional Asian market, where skinned fruit bats are sold for food. Strains of viruses that affect people, such as SARS, Ebola, Marburg and Covid-19, have been linked to bats.**

change is as serious a crisis as Covid-19. China had the highest percentage – at over 80 per cent – which bodes well for the planet (although, worryingly, the lowest score came from the USA, at just under 60 per cent). But, when we emerge from the pandemic, will these feelings usher in a greener economy?

The worldwide lockdown has had a particularly dramatic effect on international and regional transport. Air travel was down to around 10 per cent of what it was pre-lockdown, and road travel – in the UK at least – was at approximately 30 per cent of what it was BC (Before Covid). As a result, carbon emissions dropped significantly (17 per cent lower in the UK in April 2020 and up to 25 per cent lower in China during the height of the pandemic) – as did nitrogen dioxide, a polluting gas that causes, amongst other things, smog. One rather lovely consequence of this global drop in smog occurred in northern India where residents of some cities were able to see the Himalayas for the first time in more than 30 years. It was a similar story in the UK, where it's estimated that the air quality in peak lockdown was as clean as in the early 1900s. And in Venice, the near absence of boat traffic stirring up the canals has led to water so crystal clear that fish, octopuses and jellyfish could be seen. So, could this be a vision of a new future?

Of course, air and road travel bounced back to some extent when restrictions on movement were lifted, but will the Earth-friendly habits that became the norm during the pandemic continue? It's likely that large numbers of people in the UK will still want to go on overseas holidays each year, but overall business travel may never return to the level it was before – not just for financial reasons, but because internet meetings on Zoom and other forums have revealed a new and efficient way of working. Road travel – a significant factor in carbon emissions – may also not return to pre-lockdown levels as people see the benefits of home working, even if it's just for a day or two a week. And when travelling to work there's a good chance that considerably more people will either walk or cycle than did before.

Covid-19 has given us time to reflect and, with nowhere else to go, time to discover (or rediscover) the great outdoors and nature. There may be serious economic challenges ahead after months of huge government borrowing, but there's a growing feeling that now, more than ever, we need to build a better future – one that is based on sustainable energy, which, at the same time, would turbocharge a reduction in carbon emissions. Indeed, some countries see this as a way to kick-start their economies, while, at the same

Below **On a beach in São Miguel do Gostoso, northeast Brazil, children play near an onshore wind farm.**

time, boosting employment with large new, green infrastructure projects. The European Union, for instance, is planning to spend billions of Euros on decarbonising its members' economies and making homes more efficient. The British government seems to be of a similar mind and has just given the go-ahead to the largest solar plant in the UK which, when complete, will deliver clean renewable energy to over 90,000 homes.

The pandemic has both highlighted our shortcomings and shown how important it is to make us more resilient to climate change by investing in clean technologies. The visionary leaders will be the ones who say we can't afford *not* to move towards a sustainable, greener economy.

INDEX

Note: page numbers in **bold** refer to information contained in captions.

acacia 285
acidification 295
adaptive radiation 230–1, **231**
Aeberhard, Matt 100–1, **254**, 256–7, **256**
aestivation 107
Africa 82, **102**, 124, 237, 246, 251–3, **251**, 280, 282, **282**, 296, **298**, 310
agriculture 288, 289, **289**, 304
Agulhas Current 156
air travel 315
Alaska 74–5, **74**, 144
Alberta, Canada **58**, 278
Aldabra **140**, 232–4, **232**, **304**
algae 33, 150, **162**, 252
   blue-green 182, 252
Algeria **311**
Alvish, Milene **291**
Amazon rainforest 96, 100–2, 286–91, **286**
   deforestation 288–9, **288**, **289**
   fertility 134
   flooding 88–9, **88–9**, 91–5, **91**
Amazon River 88, 147
   tributaries 93, 97, 100
Amazon Tall Tower Observatory 309, **309**
Amboseli National Park **264**
Ambrym, Vanuatu 259, **259**
America
   Central 20
   East Coast 206
   North **44**, 73, 237
   South 20, 73, 237
   see also United States
American Midwest 268, 286
American Samoa **218**
American South 144
amphipods **150**
Anavilhanas Archipelago forest **91**
anchovies 154, 302
Andalusia, Spain 140
Anderson, Doug **192**
Andorra 269
Antarctic 72, 102, 108, 237, 259, **273**, 302–3
Anthropocene 276
antifreeze, natural 45
ants 24
   colonial 89
   fire **88–9**, 89, 91–5, **91–2**
   silver 68, **68**, 70, **70**
   yellow crazy 120

Arabian Gulf **149**
Arctic **51**, 72, 108, 144, **179**, 272, 273, 302
   Canadian 36–40, **36**, **38**
Arctic Ocean 272
Argentina 268, 286
`Ark of Life' 312
Asia 108, 237, 244, 296
Asian monsoon 116, 169
Asian Plate 237
Asphyxia, Mount **215**
Assumption Island 232
Atacama desert 102, 134
Atkinson, Ryan 120, **123**
Atlantic Ocean 78, 88, 132, 134, 144, 146, 149, **152**, 169, **215**, 268
atmosphere 212, 277
atolls **140**, 232–4, **232**
Attenborough, Sir David 117, 118–19, 232
Australasia 237
Australia 144, **198–9**, 199, 251, 268–9, 271, **271–2**
Australian monsoon 116, 169
autumn 64–6, **65**
avocets 182

bacteria
   halophile 257
   marine 150
Badlands National Park 79, **134**
Bahamas 180, **186**, 189, **189**, 190, **190**, **192**
baitballs 154–7, **156–7**, 199
Bali 206
Balnakeil Bay, Scotland **174**
Banaue, Luzon Island **86**
Banda, Juan Carlo 165
baobab 285
Bárcena volcano **209**
barnacles 182
   goose **182**
bats 24, 254, 314
   fruit 82–4, **82**, **84**, **314**
Bay of Fundy, Canada 175
Beacham, Dan 172–3
bears 44
   black 138
   brown 240–1, **240**, 244–6, **246**, **249**
   sun 24
Bebedo, Mozambique **269**
bee-eaters, carmine 124–31, **127**, **130–1**
Beecham, Daniel **173**
beetles, Tok Tokkie 107
Belgium 269
Bering Strait 144
Bermuda Triangle 149
Big Island, Hawaii 218
Bijagós Islands 188

Bimini, Bahamas **186**, 190, **190**, **192**
biodiversity 27, 33, 160, 182, 230, 236–7, 244, 276–7, 286, 288, 290, 295
birds 276
   see also specific species
blackbirds, red-winged 58
Blue Planet (TV series) 28
Bob Barker (ship) 296, **298**, 299–301, **299**
Bolt, Usain 70
Bonaire, Caribbean 33
boobies, Nazca 230–1, **230–1**
Borneo **20**, 169
boxfish 169
brachiating **24**
brackish water 188
Brake, Duncan **190**
Brazil 91, 100, 288–9, **288–9**, 309, **309**, 316
Britain 78–9, 237, **276**, 317
   see also United Kingdom
Brown, John 93, 94–5
Brown, Leslie 254, 256
bushfires 269, **271**, 272
Bwenge **212**
bycatch 298, 301, **301**

calcium carbonate 33, 295
California **58**, 102, 269
camels 107–13
   domesticated Bactrian 108
   wild Bactrian 107–13, **108**, **110**, **113**
camouflage 42, **170**, **286**
Campbell, Minnesota **132**
Canada 36–40, **36**, **38**, 46–9, **46**, 51–7, **56**, **58**, **72**, 146, **146**, 175, 268, **278**
canopy 20
Cape Cod 306
capelin 182
caracara 99
carbon 277–8
   low-carbon cultures 311
   stores 285–6, 288, 290, 292–5
carbon cycle 277, 293
carbon dioxide 27, 147
   and coral 33
   fixation 20, 285, **295**
   and the oceans 147, 295
   and the Permian mass extinction 265
carbon dioxide emissions 277–8, **278**, 282, 290
   agreements on 308–9
   and the Covid-19 pandemic 315–16

and deforestation 288
monitoring 309, **309**
and the oceans 292
   reduction 311
and volcanoes 212, 234, 259, 277
Western 285
carbon footprints 311
carbon offsetting schemes 290
carbon sinks 273, 286, 292
carbon-free economies 309
Carboniferous era 277
Caribbean 33, 268
   see also Bahamas
carotenoids 65–6
Castle Geyser, Yellowstone **238**
casuarina 232
Cat Tien National Park **24**, 30
cattle ranching **289**, 311
Cecropia 20
cephalopods 169
Chaitén volcano 262
Charles, Ed 84, 113, 164, 190, **190**, 192, 201
cheetahs 246
Chiang Mai **31**
Chile 102, 134, 160, 262
China 66, **66**, 108, 134, 206, 296, 310, 315
chlorophyll 65, 150
chloroplasts 20
Christmas Island 116–23, **116–18**, **120**, **123**
cinder cones 215
circadian clocks 182
circadian rhythms 58
circular economies 311
clams 182, 295
climate 79, 80, 116, 124, 273
   see also weather
climate change 101, 150, 215, 234, 278, 280, 282–3, **285**, 291, 303, 308, 314–15
cloud forests **286**
clouds 86, 88, 102
   mammatus **134**
   volcanic ash 206, **206**, 244
coal 277
cockles 182
cod 306
Colombia 209
condensation 86
Congo, Democratic Republic of 124, 259
Cook, Captain James 260
copepods **152**
Copetti, Mauricio **101**
coral 295
   deep-sea 33
   mushroom **33**
   scroll **33**
coral bleaching **292**, 295, **295**
coral polyps 33

coral reefs 33, **33**, 232, **292**, 295
Coral Triangle 169–70
Coriolis Effect 132, 147
cormorants 160–2, **160**
Cosanga, Napo Province **286**
Covid-19 pandemic 314–17, **314**
cows 138, 139
coyotes 240
crabs 176
   blue 306
   larvae 119, **152**
   orangutan 172
   red 116–23, **116–18**, 120, **123**
   robber 118, 120–3
   Sally Lightfoot **160**
Crested Butte, Colorado **65**
Cretaceous period 264, 265
crocodiles 84, 127, 130–1, **131**
Cromwell Current 160
crop failure 206, 268, 283, 285, **285**
Csaj, Lake **79**
ctenophores 152
cuttlefish, flamboyant 169–73, **169**, **170**, **172–3**
Cyclone Idai 268, **268–9**
cyclones 139, 268, **268**, **269**

Darlington, Sophie 120–3
Darwin, Charles 232
Davis, California **58**
day length 18, 20, 58
   see also photoperiodism
De Roy, Tui 228
de Vos, Asha 296, 303, 312
Death Valley, California 102
decarbonisation 317
Deccan Traps, India 264–5
deforestation 100, 101, 288–90, **288**
deoxyribonucleic acid (DNA), preservation 312, **312**
depth of field 28
desertification 282, 285
deserts 68, **68**, 102–10, **108**, **110**, 113, 134, 285, **311**
   coastal 102
   cold 102, 108, **113**
   subtropical 102
diatoms 150, **150**
dinoflagellates 150
dinosaurs 264–5, 312
diurnal animals 58
Djibouti 285
dogs, wild 246

dolphins 154, 155, **157**, 296, 298
  bottlenose 180, **180–1**
  common **156**, 157
  long-beaked common **154**
  river (boto) 88
  super pods **154**, 156, 157
dormancy 66
drones **84**, 113, 157, **179**, 227, **227**
drought **105**, 132, 134, 280, 282–3
dry seasons 96, 99, **99**, 101, 124, **124**, 127, **127**, 280
ducks, eider 176, **176**, 179, **179**
Dunes of Sossusvlei **104**
dunlins 182

EAC *see* East Australian Current
eagles
  fish 127, **127**
  Philippine **86**
Earth
  asteroid collisions 174, 264–5
  atmosphere 212, 215
  core 215, 218, 244, 246
  crust 215, 218, 236
  as Goldilocks planet 15
  magnetic field 215
  mantle 215, 236–7
  molten interior 215, 236, 246
  and the moon 174–5
  orbit 18
  rotation 86, 88, 132, 147, 174
  tilt 18, 36, 44, 72, 86
earthquakes 227, 237, 251, 259, 262
East Africa Rift 251
East Australian Current (EAC) 149
East Gippsland, Australia **271**
Ebola 314, **314**
echolocation 180, **181**
Ecuador **286**
eels
  European 149–50, **149**
  garden 180
Egypt 310
Egyptian empire 134
elephants **124**
  African **138**, **264**, 280, **280**
  desert 139
Ellesmere Island 36–7, **36**, 38–43, **38**
end-Permian extinction 264–5
endangered species 296

Endeavor (space shuttle) **212**
energy, renewable 308–10, **308**, **310–11**, 317
equator 20, 68, 102, 132, 134, 160, 271
estuaries 182, 186
Ethiopia 259
Etna, Mount, Sicily **262**
eucalyptus 271, **271**, **272**
Europe 146, 206, 237, 269, 296
European Union (EU) 317
evaporation 86
evolution 237
  iterative 232
Eyjafjallajökull volcano **206**, 209

Falkland Islands 201, **201**
feeding frenzies 154–7
Fernandina Island 160–4, **160**, **164**, **166**, 222–9, **222**, **225**, **227–9**, 254–7
fertilisers 304
fig 24–7, **24**, **27**, **30–1**
  strangler **30**
fig wasp 24–8, **27**, 31
fig wasp parasite 27
film lights 28
finches
  Galápagos 230–1
  sharp-billed ground 230
  vampire 230–1, **231**
fires
  of deforestation 101, **288**, 289, **289**
  wild 269, 271, **271**, **272**, 277, 282
fish 73, 88, 89, **91**, 144, **198**, 201, 276, **302**
  see also specific fish
fishing 296–301, **296**, **298**, **301**
fishing nets, purse seine 296, **296**, 301
fissures, radial **227**
flamingos, lesser **236**, 251–4, **251–4**, 256–7
flooding 88–95, **88–9**, **91–2**, 99–101, **99**, 132, 134, 268, 271, 282
  tidal 186, 188
fluorescence 33
fog 104–5
food chains 293
food supplies 73–5, 82
food webs 154, 188
forest recyclers 117
forests 66, 82–4, **84**
  as carbon stores 286
  cloud **286**
  eucalyptus 271, **271**, **272**
  pine 232
  regeneration projects 290–1, **290**, **291**
  saltwater **182**, 186–92

tropical 20–31, **20**
  see also jungles; rainforests
fossil fuels, burning 277, 308, 311
fox, Arctic **51**, 52–3, **52–3**, 56–7, **56**
France **149**, **152**, 269
Franke, Emily 299, 301
French Polynesia **295**
freshwater 86, 88, 139, 186, 286, 302
Friendly Floatees (rubber ducks) 144, 146
frigate birds, great **209**
frogfish 169
frogs
  Amazonian horned **286**
  Australian burrowing **105**
  desert rain 102–7, **102**
  rain 102–7, **102**
  wood 44–5, **44**
frostbite 37
Frozen Zoo 312, **312**
frugivores (fruit-eaters) 82–3
fumaroles 238

Gabon 296–8, 299, 301
Galápagos Islands 160–7, **160**, **164**, 218, 222–31, **222**, **225**, **227–30**, 254–7
Gallardo, Rafael 165
gannets 154, 155
  Cape **157**
gas 277
geese, snow 51–3, **51**, **52–3**, **56**, 57
Gelantipy **271**
geothermal areas 238–43, **240**, **242–3**
Germany 269
gers 110
geysers 238, **238**, 240–1, **240**
gibbons 24
  southern yellow-cheeked crested **24**
glacial melting 303
Global Airborne Observatory 290
global conveyor belt 147, 302, 303
global warming 234, 271–3, 276, 278, **278**, 280, 282–3, 290, 293, 303–4, 306, 308–9
Gobi Desert 102, 107–10, **108**, **110**, 113, **113**
Gondwana 237
gorillas, silverback mountain **212**
Goulokum Tegg 283, 285, **285**
Grand Terre, Aldabra 234
Great Barrier Reef **292**, 295

Great Gobi National Park 110
Great Green Wall **282**, 285
Great Plains 132
Great Rift Valley 246
greenhouse gases 212, 278, 285, 288, 311
Greenland ice sheet 302
grunions 182
Gulf of Maine 306
Gulf of Mexico 78, 146, 306
Gulf of St Lawrence **146**
Gulf Stream 78, 146, 149–50, 303
Gulf of Thailand **302**, 304
gulls 199

habitat destruction 278
Hadley, George 102
Hadley Cells 102
halibut 296
Hammarstedt, Captain Peter 299
hamsters, Siberian 58
hardyhead **198**, 199, **199**
hares, Arctic 42–3, **42**
Hawaii 144, **211**, 215, 218, **218**, 259
hawks, Swainson's 73
hearing 138–9
heatwaves 269, 271, 282
helium 14–15
Heron Island **292**
herons 199
  grey **79**
herring 154
hibernaculum (shallow burrow) 44
hibernation 44, 46, 66
Himalayas 108, 237, 315
Holocene 124, 132, 265, 276
home working 316
Hong Kong 144
hornbills 24
  rhinoceros **24**
Horrocks, Roger 155–6
hot springs 238
Houck, Marlys 312, **312**
human impact 266–317
human population growth 276, 308
Humboldt, Alexander von 160
Humboldt Current 160
humidity 117
Hungary **79**
hunting 99, 101, 130
Hurricane Katrina 175
Hurricane Sandy 132
hurricanes 132, 134, 175, 268, 271, 282
hydrogen 14–15
hyena 246, 252

ice ages 276
ice sheets 234, 302–3
iguana
  land 222–6, **222**, **225**, **228**, 229

marine 160–7, **160**, **162**, **164**, **166**
Ilyinsky volcano **249**
India 116, 237, 264–5, 310, 315
Indian Ocean 132, 169, 232, **232**, **304**
indigenous peoples 289, 290–1
Indonesia **33**, **86**, 134, 144, 172, **172–3**, 204, 206, **206**
Indonesian Archipelago 169
Indonesian Through-Flow (ITF) 169
infrasonic sound 138–9
International Space Station (ISS) **146**
intertidal pools 182
invasive species 120
Iquitos, Peru 92–5
Iran 268
Isabela Island 229
Italy 209, 262

jackals 252
jaguars 88–9
Japan 74–5, 79, 144, 268
Java, Indonesia 134
jellyfish **150**
jet lag 58
jungles 20–31, **20**, 68, 124, 288, **290**, 291

Kamchatka Peninsula 74–5, **212**, **236**, 237, 240–3, **240**, **242–3**
kangaroo rats 107
kangaroos, eastern grey **271**
Karan Island **149**
Karrak Lake 51–7, **56**
Kasanka National Park 82–4, **82**, **84**
Kenya **80**, **138**, **264**, 280, **280**, **312**
keystone species 24
Kīlauea volcano **211**, 218, **218**
Kilimanjaro, Mount **264**
Kirby, Richard 70, **70**
Kiskunsági National Park **79**
Kliuchevsko volcano **212**
knots 182
koalas **271**
Komodo Island **33**
Koyli Alpha, Senegal **283**, **285**
Krakatoa 205, 209
Kreisel tanks 152
krill 73, **74**, 75, 201
Kurile Lake 244–5, **244**, **246**, **249**

La Cumbre volcano 222–9, **222**, **225**, **227–9**
land degradation 282, 285
land formation 218, **218**

ACKNOWLEDGEMENTS

Neither the book nor the TV series would exist without the generous help of scientists, wildlife researchers, conservation and wildlife organisations, universities and local wildlife experts. Ultimately, it is these people and organisations that we rely on so much for these shows.

For turning these stories into compelling sequences, I would like to thank the excellent Silverback Films production team on *A Perfect Planet* for their dedication, passion and hard work during the four years it took to make the series.

Behind the cameras and behind the scenes are the camera crews, sound designers, editors, edit assistants and technicians, who make these stories look and sound so good. In this business we are constantly required to raise the bar on what has gone before, and all those involved in *A Perfect Planet* did just that.

I would also like to thank my wonderful family – my wife, Lizzie, and children, Jonjo, Jules and Leila – for their patience and understanding during my regular trips overseas. Despite the incredible experiences I've been extremely fortunate enough to have had over the years, travelling around the world, I've always found time to miss them (though I'm sure they were also occasionally relieved to have had a break from my mild OCD with tidiness!).

Finally, for this stunning book that complements the series so well, I would very much like to thank Alastair Fothergill – a legend in the natural history film business – for his perfect foreword, and the very supportive team at Penguin Random House, in particular Nell Warner, Michael Bright, Steve Tribe and last, but certainly not least, Laura Barwick, who was responsible for compiling the fantastic photos in this book.

PRODUCTION TEAM

Sir David Attenborough

Tom McDonald
Jack Bootle

Alastair Fothergill
Anna Findlay
Anna Kington
Dan Clamp
Daniel Rasmussen
Darren Clementson
Darren Williams
Ed Charles
Eleanor Perryman
Ellie de Cent
Elly Salisbury
Emily Franke
Gina Shepperd
Huw Cordey
Jane Hamlin
Jenni Collie
Jonnie Hughes
Judi Oborne
Keith Scholey
Lauren Childs
Nick Shoolingin-Jordan
Rachel James
Rob Childs
Rosie Lewis
Sarah Garner
Sarah-Jane Walsh
Sean Pearce
Tash Dummelow
Toby Nowlan
Vicky Singer

FIELD CREW
Basti Hofmann
Daniel Beecham
Denis Mollel

Jose Masaquisa
Juan Carlos Balda
Matt Carr
Moses Mdama
Rafael Gallardo
Thomas Boyer
Tui de Roy
Waldo Etherington
Zheng Yi

CAMERA AND SOUND TEAM
Alain Lusignan
Alastair MacEwen
Alexander Vail
Barnaby Trevelyan-Johnson
Barrie Britton
Bernt Bruns
Bertie Gregory
Braydon Maloney
Chad Cowan
Darren Williams
Dawson Dunning
Doug Anderson
Duncan Brake
Dustin Farrell
Guillermo Armero
Hal Smith
Howard Bourne
Hugh Miller
Ivo Nörenburg
Jacky Poon
Jesse Wilkinson
John Aitchison
John Brown
John Shier
Johnny Rogers
Jon Krosby
Julius Brighton
Justin Maguire
Kieran O'Donovan

Lianne Steenkamp
Marcus Coyle
Mark MacEwen
Matt Aeberhard
Mike Guarino
Mike Olbinski
Nick Shoolingin-Jordan
Olly Meacock
Oliver Richards
Parker Brown
Paul Klaver
Paul Thompson
Paul Williams
Peter Elliot
Richard Kirby
Richard Wollocombe
Roger Horrocks
Rolf Steinmann
Ryan Atkinson
Sam Stewart
Sean Millar
Simon de Glanville
Sophie Darlington
Steve Brooks
Tamara Stubbs
Tom Beldam
Tom Crowley
Tom Rowland
Tom Varley
Tom Walker
Will Steenkamp
Zubin Sarosh

POST PRODUCTION
Films at 59
Miles Hall

MUSIC
Ilan Eshkeri
Steve McLaughlin

FILM EDITORS
Charles Dyer
Darren Flaxstone
Dave Pearce
Emily Davies
Jacob Parish
Nigel Buck
Sam Rogers

ONLINE EDITORS
Franz Ketterer
Wesley Hibberd

SOUND EDITOR
George Fry

DUBBING MIXER
Graham Wild

COLOURIST
Adam Inglis

GRAPHIC DESIGN
Moonraker
Contains modified Copernicus Sentinel Data 2020

FOLEY
Andy Devine
Brian Moseley
David Yapp
Richard Hinton

VFX
Chris Gunningham
Shaun Littlewood

BBC STUDIOS DISTRIBUTION
Mark Reynolds
Monica Hayes
Patricia Fearnley

CO-PRODUCERS
China Media Group CCTV9
Discovery
France Télévisions
Tencent Penguin Pictures
The Open University
ZDF

ACADEMIC CONSULTANTS FOR THE OPEN UNIVERSITY
Dr Philip Wheeler
Professor Stephen Lewis

WITH THANKS TO
Cristina Hughes
Helen Healy
NASA

VOLCANOES
Flying Medical Service
Galápagos National Park
Keira Malik
Martha Masden
Reno Sommerhalder
Richard & Janice Beatty
Roberto Pepolas
Seychelles Island Foundation
Simon Villamir
Tanzania National Parks
The British Hovercraft Company
Ulla Lohmann

University of Bristol
Vanuatu National Cultural Council
Yellowstone Expeditions
Yellowstone National Park

THE SUN
Ahiak (Queen Maud Gulf) Migratory Bird Sanctuary
Aziz Kheraj
Dr Bach Thanh Hai
Cat Tien National Park
Centre Cinématographique Marocain
Dr Clara do Amaral
Dan MacNulty
Dana Kellett
Ellen Xu
Emily Jayne Turner
Dr Gustaf Samelius
Janice Straley
Karrak Lake Research Station
Kayla Buhler
Madison Kosma
Manitoba Sustainable Development
Marina Kenyon
Matthew Chippendale-Jones
Mount St. Joseph University
Fan Penglai
Pauline Bloom
Plitvice Lakes National Park, Croatia
Queen Sirikit Botanical Gardens
Dr Ray Alisauskas
Serge Aron
Shennongjia National Park and Science Institute
Dr Stephen Compton
Dr Wattana Tanming
Younes Ounaceur

WEATHER
A–Z Solutions Ltd
Adiya Yadasuren
Alan Channing
Amy Thompson
Carolina Ribas and Flavia Rocha
Christmas Island National Park
Claire Powell
Colin Beale
David Maguire
Dick Vogt
DNPW / Ministry of Tourism Zambia
Etty Bartenshtein
Frank Willems
Great Gobi A Special Protected Area's Administration
Ian & Lesley Thomson
Jody Taft
John Hare
Kasanka Trust ltd
Kenneth Butler
Khishigjargal Dorjjugder

Lisa Connaire
Mauricio Copetti
Ministry of Foreign Affairs of Mongolia, Department of Public Diplomacy and Cultural Cooperation
Panoramic Journeys
Richard & Janice Beatty
Rob Muller
Sean O'Donnell
Shenton Safaris – Kaingo – South Luangwa National Park
Tapiche Jungle Lodge
Tiwi Land Council and landowners
Wild Camel Protection Foundation
Zimbabwe Parks and Wildlife Management Authority

OCEANS
Carl Zeiss Microscopy France
Christian Sardet
Elephant Seal Research Group
Galápagos National Park
Gisle Sverdrup
Julie Hartup
Leo Leibovici
Lizard Island Research Station
Dr Lyle Vail
Mick & Yuta O'Shea
National Department of Environment Affairs, South Africa
NAD Lembeh
Neal Watson
Nelson Mandela Bay Municipality
Noé Sardet
Parafilms
Pt Lahuka Indonesia Ekspedisi
Raggy Charters
Sea Lion Lodge
Sharif Mirshak
Simon Buxton
Simon Villamir
Thai Film Office
Tom Morris
UPMC – Université Pierre et Marie Curie – CNRS Observatoire Océanologique de Villefranche sur Mer
Vegard Sandvik

HUMANS
Amazon Tall Tower Observatory
Angela Sheldrick
Anjali Raman-Middleton
Dr Asha de Vos
Ava Crofts
Bob Prescott
Celina Maria Müller Ferreira Pinagé
Charles Dyer
David Sheldrick Wildlife Trust
Dionne Scougul

Direction Générale des Pêches et de l'Aquaculture, Gabon
Dulce Continentino
Flavio Somogyi
Great Green Wall
Henry Scott
Instituto Brasileiro do Meio Ambiente e dos Recursos Naturais Renováveis, IBAMA
Instituto de Proteção Ambiental do Amazonas, IPAAM
Izzy Warren
Jamie Unwin
Jeremy Rifkin
Josh Forwood
Korka Ba
Marine Nationale, Gabon
Marlys Houck
Marshfield Airport KGHG
Mass Audubon Wellfleet Bay Wildlife Sanctuary
Max Smith
Milene Alves Oliveira
National Department of Environmental Affairs, South Africa
New England Aquarium
Dr Niall McCann
The Olde House, Cornwall
Oliver Crofts
Parafilms
Christian Sardet
UPMC – Université Pierre et Marie Curie
Peter Hammarstedt
Rede de Sementes do Xingu/ Instituto Socioambiental (ISA)
RZSS Edinburgh Zoo
Special thanks to San Diego Zoo Global
Santiago Botia B
Scarlett Westbrook
Sophia Dyvik-Henke
Theodore Siri
Tim Lenton
Tiwi Land Council and Landowners
Turtles Fly Too
UK Student Climate Network

1

BBC Books, an imprint of Ebury Publishing
20 Vauxhall Bridge Road, London
SW1V 2SA

BBC Books is part of the Penguin Random House group of companies whose addresses can be found at global.penguinrandomhouse.com

Penguin Random House UK

Copyright © Huw Cordey 2020

Huw Cordey has asserted his right to be identified as the author of this Work in accordance with the Copyright, Designs and Patents Act 1988

This book is published to accompany the television series entitled *A Perfect Planet*, first broadcast on BBC One in 2021.

Executive producer: Alastair Fothergill
Series producer: Huw Cordey

First published by BBC Books in 2020

www.penguin.co.uk

A CIP catalogue record for this book is available from the British Library

978-1-785-94529-8

Publishing Director: Albert DePetrillo
Project Editor: Nell Warner
Picture Research: Laura Barwick
Image Grading: Stephen Johnson, www.copyrightimage.co.uk
Design: Bobby Birchall, Bobby&Co
Production: Antony Heller

Printed and bound in Italy by Printer Trento

Penguin Random House is committed to a sustainable future for our business, our readers and our planet. This book is made from Forest Stewardship Council® certified paper.

PICTURE CREDITS

Front artwork © BBC Studios Distribution 2020. Photography © Getty Images.
Back photography by Toby Nowlan © Silverback Films.
Photo on back flap © Huw Cordey

1 Jack Dykinga/naturepl.com; **2-3** Federico Veronesi; **4-5** Alex Mustard; **6l** New Zealand American Submarine Ring of Fire 2007 Exploration, NOAA Vents Program, the Institute of Geological & Nuclear Sciences and NOAA-OE/SPL; **6r** Dr Ken Macdonald/SPL; **7** Guy Edwardes/naturepl.com; **8-9** Thomas Mangelsen/Minden/naturepl.com; **10** Ashley Cooper/naturepl.com; **11** Huw Cordey

The Sun
12-13 Alain Lusignan; **14-15** Smileus Images/Alamy; **16-17** Brandon Cole; **18-19** Tetra Images/Alamy; **21** Huw Cordey; **22-3** Daniel Salgado; **24** Sarah-Jane Walsh; **25** FLPA/Alamy; **26-7** Alastair MacEwen/Silverback Films; **29** Nick Shoolingin-Jordan; **30** Jonnie Hughes; **31** Nick Shoolingin-Jordan; **32** Danita Delimont Creative/Alamy; **33** Ethan Daniels/Alamy; **34-5** Alex Mustard; **36-7** Staffan Widstrand/naturepl.com; **37** Jesse Wilkinson/Silverback Films; **38-9** Alain Lusignan; **40** Alain Lusignan; **41t** Alain Lusignan; **41m, b** Rolf Steinmann/Silverback Films; **42-3** Rolf Steinmann; **42-3b** Rolf Steinmann/Silverback Films; **44-5** Richard Kirby/Silverback Films; **46** Nick Shoolingin-Jordan/Silverback Films; **47** Nick Shoolingin-Jordan; **48-9** Huw Cordey; **50-1** Sergey Gorshkov/naturepl.com; **52-3** Sergey Gorshkov/Minden/naturepl.com; **54-5** Sarah-Jane Walsh; **56-7** Ivo Norenberg; **59** Minden/Alamy; **60-1** Neil Losin; **62-3** Richard Kirby/naturepl.com; **64-5** rapp/123RF; **66** Jacky Poon/Silverback Films; **67** Ellen Xu; **69** Richard Kirby/Silverback Films; **71** Nick Shoolingin-Jordan; **72-3** All Canada Photos/Alamy; **74-5** Nick Shoolingin-Jordan (National Marine Fisheries Service MMP A/ESA Permit No. 18529)

Weather
76-7 Cavan Images/Alamy; **78-9** Bence Mate/naturepl.com; **80-1** Denis-Huot/naturepl.com; **82-3** Ed Charles; **85t** Ed Charles; **85b** Nick Garbutt; **86** Thomas Marent/Minden/naturepl.com; **87** Image Professionals GmbH/Alamy; **88** John Brown/Silverback Films; **89l** Alex Vail/Silverback Films; **89r** John Brown/Silverback Films; **90** Minden/Alamy; **91t** John Brown/Silverback Films; **91b** Alex Vail/Silverback Films; **92-3** John Brown/Silverback Films; **94l** Toby Nowlan; **94r** Alex Vail; **95** Mark Moffett/Minden/naturepl.com; **96-7** Darren Williams; **98t** Darren Williams/Silverback Films; **98b** Darren Williams; **99l** Darren Williams; **99r** Tom Crowley/Silverback Films; **100** Tom Rowland; **101** Darren Williams; **103** Justin Maguire/Silverback Films; **104-5** Hougaard Malan/naturepl.com; **106-7** Avalon/Photoshot/Alamy; **108-9** Ed Charles; **110** John Shier/Silverback Films; **111** Ed Charles; **112t** Ed Charles; **112b** Huw Cordey; **114-5** Ed Charles; **116** ATARA Film/Moritz Katz; **117** Sophie Darlington/Ryan Atkinson/Braydon Moloney/Silverback Films; **118-9** Amy Thompson; **121** Braydon Moloney; **122** Amy Thompson; **123** Ingo Arndt/naturepl.com; **124** Bernt Bruns/Silverback Films; **125** Paul Klaver/Silverback Films; **126t** Will Burrard-Lucas/naturepl.com; **126b** Toby Nowlan; **127** John Aitchison/Silverback Films; **128-9** Toby Nowlan; **130-1** Tom Rowland/Silverback Films; **133** Chad Cowan; **134-5** Cavan Images/Alamy; **136-7** Chad Cowan; **138-9** Denis-Huot/naturepl.com; **140** Chris Jobs/Alamy; **141** Cheryl-Samantha Owen/naturepl.com

Oceans
142-3 Norbert Wu/Minden/naturepl.com; **145** Paulo Oliveira/Alamy; **146** NASA/SPL; **147** Moonraker/Silverback Films; **148** Michael Pitts/naturepl.com; **149** BIOSPHOTO/Alamy; **150** Alastair MacEwen/Silverback Films; **151** Greg Lecoeur; **153tl** Daniel Rasmussen; **153** Solvin Zankl/naturepl.com; **154-9** Greg Lecoeur; **160** Tui De Roy/RovingTortoisePhotos; **161** Ed Charles; **162-3** Tui De Roy/RovingTortoisePhotos; **164** Richard Wollocombe/Silverback Films; **164-5** Tui De Roy/RovingTortoisePhotos; **165** Joe Stevens; **166-7** Tui De Roy/RovingTortoisePhotos; **168** Alex Mustard; **169** Hugh Miller/Silverback Films; **170** Hugh Miller/Silverback Films; **171** Alex Mustard; **172-3** Daniel Rasmussen; **174-5** Fortunato Gatto; **176** Jon Krosby/Silverback Films; **177t** WildWonderofEurope/Lundgren/naturepl.com; **177b** Doug Anderson/Silverback Films; **178** Daniel Rasmussen; **179** Tom Rowland; **180-1** Duncan Brake/Silverback Films; **183** Konrad Wothe/Minden/naturepl.com; **184-5** Paul Marriott/Alamy; **186-7** Ed Charles; **188** Pedro Narra/naturepl.com; **189** Doug Perrine/naturepl.com; **190-3** Ed Charles; **194-5** Daniel Rasmussen; **194-5b** Roger Horrocks/Silverback Films; **196-7** Julie Hartup; **198-9** Alex Vail/Silverback Films; **200** Tui De Roy/RovingTortoisePhotos; **201** Sarah-Jane Walsh

Volcanoes
202-3 Joep Rutgers/Alamy; **204-5** AirPano LLC; **206** Martin Rietze/SPL; **207** Arctic Images/Alamy; **208-9** Claudio Contreras/naturepl.com; **210-1** Doug Perrine/naturepl.com; **212** NASA; **213** Chris Whittier; **214** Huw Cordey; **216-7** MZPHOTO.CZ/shutterstock; **218-9** Floris van Breugel/naturepl.com; **220-1** Doug Perrine/naturepl.com; **222-4** Tui De Roy; **225** Sam Stewart/Silverback Films; **226-8** Tui De Roy; **229** Sam Stewart/Silverback Films; **230-1** Huw Cordey; **233t** Moonraker/Silverback Films; **233b** Huw Cordey; **234-5** Huw Cordey; **236-7** Sergey Gorshkov/naturepl.com; **238** Jeff Vanuga/naturepl.com; **238-9** Danny Green/naturepl.com; **240-3** Huw Cordey; **244-5** Toby Nowlan; **247** Tom Walker; **248-9** Toby Nowlan; **250-2** Darren Williams/Silverback Films; **253** Matt Aeberhard/Silverback Films; **254-7** Darren Williams; **258-261** Huw Cordey; **263** Andrea Savoca/Getty; **264-5** Federico Veronesi

Humans
266-7 Nick Shoolingin-Jordan (with thanks to David Sheldrick Wildlife Trust); **268** Claudia Weinmann/Alamy; **269** US Air Force Photo/Alamy; **270** Jo-Anne McArthur/We Animals/naturepl.com; **271** Doug Gimesy/naturepl.com; **272-3** David Gallan/naturepl.com; **274-5** Florian Ledoux; **276-7** Nature Picture Library/Alamy; **278** Nick Shoolingin-Jordan; **279** Ashley Cooper/naturepl.com; **281** Nick Shoolingin-Jordan (with thanks to David Sheldrick Wildlife Trust); **282** UNCCD/Global Mechanism; **283-5** Nick Shoolingin-Jordan; **286** Alex Hyde; **287** Lucas Bustamante/naturepl.com; **288** NASA/SPL; **289** Jacques Jangoux/SPL; **290l** Parker Brown/Silverback Films; **290r** Daniel Rasmussen; **291** Julius Brighton/Silverback Films; **292-3** Gary Bell/Oceanwide/naturepl.com; **294** Alastair MacEwen/Silverback Films; **295** Alexis Rosenfeld/Getty; **297** Alex Hofford/Greenpeace; **298** Tamara Stubbs; **299** Emily Franke; **300** Tamara Stubbs; **302-3** Tony Wu/naturepl.com; **304** Huw Cordey; **305** Florian Ledoux; **306-7** Nick Shoolingin-Jordan; **308** Xinhua/Alamy; **309** Parker Brown/Silverback Films; **310** Jerónimo Alba/Alamy; **311** robertharding/Alamy; **312** Daniel Rasmussen; **313** Mark Carwardine/naturepl.com; **314** Maurizio Biso/Alamy; **315** Andrea Pattaro/Getty; **316-7** Pulsar Images/Alamy

Endpaper front Genevieve Vallee/Alamy; **endpaper back** Tui De Roy